Instructional Guide for The ArcGIS Book

Kathryn Keranen and Lyn Malone

Esri Press
Redlands, California

All images courtesy of Esri except as noted

On the cover:
The Stamen Design Watercolor Map applies raster effect area washes and organic edges over a paper texture to create an effect reminiscent of hand-drawn maps. Created with OpenStreetMap data and published as a map service for use as an alternative basemap in ArcGIS Online.

To our students past, present, and future:
You have made us laugh, you have made us
cry, but most of all you have kept us young.

Kathryn and Lyn

Table of Contents

How to use this book

This instructional guide is designed for anyone who wants to learn—or learn more—about ArcGIS Online and its supporting cast of online and mobile apps. It is also a valuable tool for those who want to help others learn. Each chapter in its earlier companion book, *The ArcGIS Book,* includes Quickstarts, with information about relevant software, data, and web resources, and Learn ArcGIS lessons to provide hands-on practice in the skills discussed. *Instructional Guide for The ArcGIS Book* starts where the first volume left off. In this guide you'll find videos and activities recommended for developing basic GIS skills and understanding, along with scenario-based lessons that put those concepts to use in answering questions and addressing real-world problems.

If a Swiss army knife is a multi-purpose, versatile, and adaptable tool, then this guide is the Swiss army knife for learning GIS. Use it as a personal tutorial, a refresher course, a lab manual, or a repository of training activities. Use it to meet your needs.

A personal tutorial

Use this guide on your own to learn about ArcGIS Online. You'll learn how to make maps and share them as web apps and story maps. You'll learn how to access existing data online and to create and map your own new data. You'll practice working with mobile apps like Survey123, Collector, and Explorer for ArcGIS. Go through the book from chapter to chapter or hone in on the topics and skills that are most relevant to your own situation. There is no wrong way to use this guide—make it work for you.

A refresher course

Use this guide on your own to update your GIS skills and knowledge. Instead of working with software and data stored on your computer, you'll learn to leverage the power of cloud computing and open data. Discover opportunities to collaborate and share your maps and data with an ArcGIS Online organizational account. Get hands-on practice in 3D GIS, live data feeds, and crowdsourcing with ArcGIS Online. Whatever your prior GIS experience, with this guide you can lift yourself to a new and higher level.

A lab manual

Use this guide with a group of students to provide practice in basic skills and content. Focus on particular chapters and skills one at a time or encourage learners to go through the entire book at their own pace. By using the book in this way, you are offering people a foundation of GIS skills and knowledge they can build upon down the road.

A repository of training activities

Use this guide to design and deliver GIS training in a focused setting such as teacher professional development. Have you been asked to provide training in "the basics" of ArcGIS Online? Look no further than the content of chapters 1 and 2. Does your audience want to learn about story maps? Use chapter 3. If they want to do their own field data collection, then focus on chapter 8. For those who are often called upon to train others in the use of GIS, *Instructional Guide for The ArcGIS Book* is a versatile tool. No matter what the focus of the training or learning, this guide provides the essential ingredients to combine and blend in any way you choose.

Rebecca Gentry teaches at Herndon High School in Farifax County, Virginia. Her Geospatial Semester Class is a dual enrollment class with James Madison University.

Introduction

This learning guide is a companion to *The ArcGIS Book: 10 Big Ideas about Applying Geography to Your World* (Esri Press, 2015); it provides further resources and hands-on lessons, to build upon the learning opportunities provided in the earlier book. You can engage with this book as a student or teacher, novice or expert, engineer or historian. Each chapter includes activities and lessons designed to take the mystery out of using geographic information system software and web technology to solve real-world problems.

This book focuses on the fluid interaction between ArcGIS Online and GIS applications on the web and on mobile devices. All the lessons in this guide can be completed with ArcGIS Online, a free resource for exploration of data and imagery. ArcGIS Online is a rich repository of maps and data resources contributed by scientists, GIS professionals, and other GIS users, including ready-made basemaps that anyone can use to ask questions and share answers.

Most of the exercises and lessons do not require logging into an organizational account. (The ones that do are color-coded purple, for convenience.) None require having data preloaded on student computers. Scenario-based, the lessons need not be done in any particular order, providing teachers with many instructional options. The authors have taught students GIS and teachers how to teach GIS for a long time. Their lessons pulse in time with the heartbeat of GIS—visualizing and analyzing spatial data—and in streamlining them, they've left nothing in the way of reaching students at their excitement level.

In a world that, in recent years like never before, is discovering big data and the power of cloud storage that makes this data accessible to everyone, there is a lot of new information waiting to be explored. Most of it is geospatial data requiring processing in a geographic information system before it can be understood and interpreted. In this era of easy access to cloud-based open data, we are all novices, all learning new things as fast as we can. Software that enables rapid analysis allows us to find order in this chaos of discovery.

So, at first glance, this is a GIS book. But once learners get their hands on the lessons within, it becomes much more. Because GIS is a learning tool itself—used to ask and answer questions—in its broadest sense, this book is nothing short of a guide in how to learn in a new era.

This book builds on the concepts and maps of *The ArcGIS Book*. Each of the chapters here is aligned with a parallel chapter in The ArcGIS Book.

Maps in that book are used as the starting point for many of the lessons here and can be accessed from the online version of *The ArcGIS Book* http://thearcgisbook.com. *The Instructional Guide for the ArcGIS Book* provides insights and instructions on how to create maps that solve problems like those presented there. This guide also includes supplemental videos, activities, strategies, and discussion questions for further exploring *The ArcGIS Book*.

Maps, the Web, and You
Take flight with ArcGIS to the cloud

GIS became available to everyone with the introduction of mapping applications like Google Earth, MapQuest, and many other apps that anyone could access on their desktop and mobile device. Now, whether you're a beginner or a professional, you can access, analyze, and share data and maps within your organization and beyond.

The activities, videos, and lessons in this chapter lead to the discovery of the uniqueness of GIS on both the web and the desktop. The five lessons in chapter 1 offer instruction in the use of three visual/analytical tools and include a scenario-based lesson on the United States population change and a how-to for adding live-feed data. This chapter ends with questions that support reading comprehension, reflection, and discussion of the narratives in the corresponding chapter 1 of *The ArcGIS Book*.

Introductory activities

Video
Videos elevate motivation and enthusiasm as well as enhance discussion. The two map videos (numbers 1 and 2) are short but compelling and could be set to loop in the background of any instructional area. Teachers run them in the background as students arrive to generate interest. Interaction with the maps results in a better understanding of the scope of work done with GIS.

To access the following videos go to video.esri.com.
Search for each video by the title.

Maps We Love, The Power of Maps (2015).

Map lovers, feast your eyes on this! 30 Years of Maps (2014).

Geography + technology = "Wow!" What Is GIS? (2014).

Activity
Explore Esri Map Books, published
since 1984:

www.esri.com/mapmuseum

GIS users around the world map mineral resources, ecosystems, hurricane surge, waterfowl migration, earthquake disasters, and wildfires, just to mention a few things. Published annually since 1984, the Esri Map Book acknowledges the important and innovative accomplishments of such users. Leafing through, you'll be amazed at the great variety of maps. As you look at the maps in the map books, select three maps and for each map record the following:

▸ Organization that produced the map

▸ Reason or problem for the map

▸ Layers included in the map

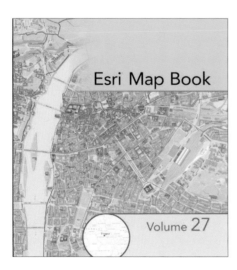
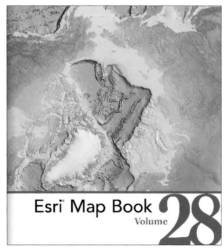
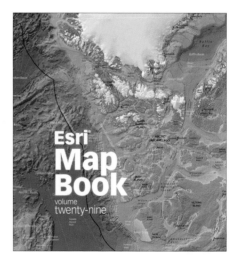

Lessons: Account not required

Remember, you can access online all you need for the lessons in this guide, either with no sign in to an ArcGIS Online organization account required (the first set of lessons in each chapter) or by signing in with an organizational account (for the last set of lessons in each chapter).

The ArcGIS Book interactive version (www. TheArcGISBook.com) starts you off by exploring many of the unique analytical maps and scenes accessible online, such as Highway Access in Europe and U.S. Populations. From the GIS map or scene interface, users take up visual and analytical tools that investigate anything from election results in 3D to the terrain of the Swiss Alps. Tools like that solve problems; they help you find answers to questions along the way; and they're available online. The following section explores three of these tools: visual recognition of spatial distribution, the analytical concepts of symbolizing/classification, and querying/filtering. Investigate each of the maps to understand the analytical value of each tool.

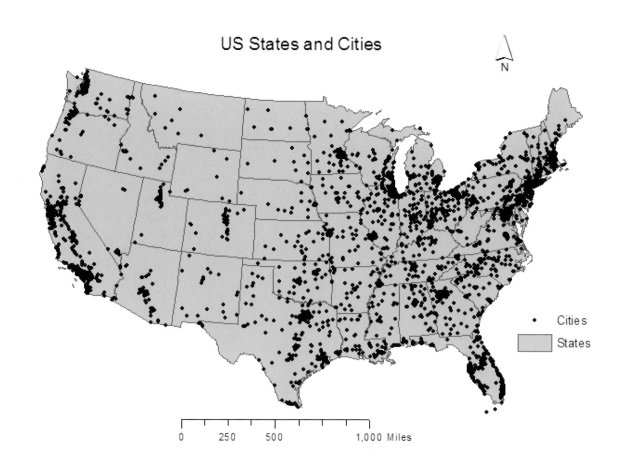

US States and Cities

N

Cities
States

| 0 | 250 | 500 | 1,000 | Miles |

Scenarios of spatial distribution

Spatial distribution is such a formal term for a skill that starts early and is basic. If you show a class of third graders the US States and Cities map, they will say things like

"There are a whole bunch of dots on the right."
"There are groups of dots here." Or,
"There aren't many here," as they point with their fingers.

All these are simple observations, easy to make, but isn't that the point? Information made visible registers instantly, especially when associated with place, which is what mapping software does. Your observations don't need to be sophisticated or use geographic terms in order to describe spatial distribution. As you advance in your ability to use such terms, however, your spatial descriptions will become more helpful to you in your collaboration and communication with other GIS users.

Looking at spatial distribution, you begin to understand why the spatial component of GIS sets it apart from other software. GIS has the element of where. Data can be displayed using many different tools, but data that has a spatial component can be displayed in maps. One of the most compelling reasons to use a GIS is that it gives the user the ability to recognize patterns from mapping. In the following exercises, lessons that employ tools within your reach, you will use maps that show patterns of spatial distribution.

Maps show the spatial distribution of phenomena and how it is arranged across the surface of the earth. The phenomena can be Earth's physical landscape, demographic patterns, economic patterns, and any other pattern with a spatial component that connects it to the earth. Spatial distribution, along with spatial analytics, is the heartbeat of any GIS system.

The maps that follow illustrate the spatial distribution of two types of demographic data and of one terrain model. Terrain models allow users to better visualize landscape. The study of population growth and trends over such landscapes allows government agencies to make better predictions as to where resources should be made available. Questions are included to help guide you in understanding spatial distribution.

Build skills in these areas

▸ Opening and modifying an existing online map

▸ Interpreting legends

▸ Determining formulas used to calculate change

▸ Writing spatial descriptions of various geographic areas. A spatial description refers to the pattern of data arranged according to its physical position or relationships

What you need

▸ Account not required

▸ Estimated time: 30 minutes for U.S. Population Change 2000 to 2010

▸ Estimated time: 15 minutes for Terrain of Swiss Alps

Lesson 1-1: Spatial distribution
U.S. Population Change 2000 to 2010

Find the USA Population change map for this lesson in the first chapter of the thearcgisbook. com. Under the ArcGIS information items heading, locate it beneath the subhead WEB MAPS and SCENES.

The "About" section of the map explains that the *U.S. Population Change 2000 to 2010* map indicates the annual rate of total population change in the United States from 2000 to 2010. The map shows two spatial components: counties and census tracts. The census tracts are not revealed until the user is zoomed in at a scale of 750 kilometers to 100 kilometers. Counties are typical political divisions. A census tract is a man-made division as well.

U.S. Population Change 2000 to 2010

Scenario

The new startup company Build for Us has had so much success that they are looking to expand. The company knows that any geographic location that is increasing in population requires more housing and would need to buy more material from Build for Us. They have asked for an analysis of counties and even sections of counties that would be a good place to locate new Build for Us facilities. They are looking for both general areas in the United States and specific areas in certain counties.

1. Click the USA Population Change Map and zoom out to see contiguous USA.
This multiscale map allows you to view geographic data across a range of scales.
   a. What is the first spatial component that is seen?

   b. What is the second spatial component that is seen as you zoom in?

   c. What does a census tract represent?

   d. Examine the legend and write a description of the legend in your own words.

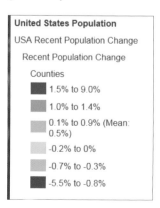

e. What formula was used to calculate the percent change in population from 2000 to 2010?

Build For Us has asked for written spatial descriptions of the United States, individual states, and specific counties. They will use these descriptions along with the map to decide where to locate new Build For Us stores.

2. Using the scalability of the map, write spatial distribution descriptions for the following areas.
   a. Write a description of the spatial distribution of the United States by county by population change from 2000 to 2010.

   b. Zoom in to your state and write a description of the spatial distribution of your state by population change from 2000 to 2010.

   c. Zoom in to your county and write a description of the spatial distribution of the census tracts by population change in your county.

   d. How could other state and county agencies use this information?

Lesson 1-2: Spatial distribution
Terrain of Swiss Alps

Find the *Terrain of Swiss Alps* map for this lesson in the first chapter of the thearcgisbook.com. Under the ArcGIS information items heading, locate it beneath the subhead Layers.

Scenario

As a new professor of geography you are looking for ways to incorporate new technologies into your classroom while still teaching core content. You have chosen to use interactive GIS software to demonstrate both physical and political boundaries. You want your students to identify the countries that are home to the Alps by looking at the spatial distribution of the terrain.

1. Click the Terrain of Swiss Alps map. It opens as a Demo Terrain map.
   a. What does a terrain map show?
2. On the top of the page, click Basemap and investigate the area using various basemap image layers.
   a. Using these basemaps, write a paragraph about the spatial distribution of the Alps.
   b. To whom would this map be interesting?

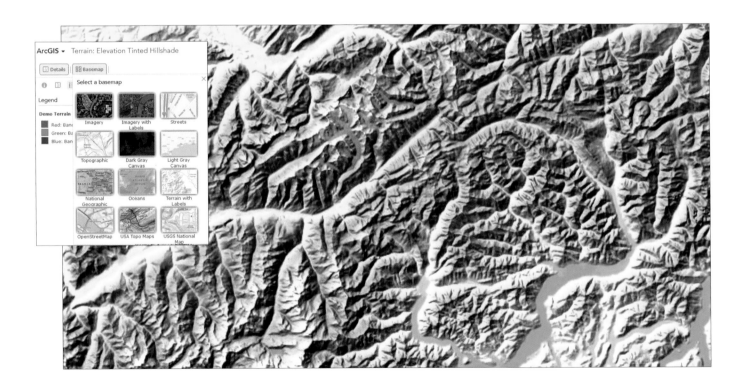

Symbolization and classification

Before explaining symbolization and classification in classical GIS terms, let's take a look at the essence of these processes.

You're familiar with using a symbol to represent something else: A heart represents love; a fist signifies anger; balloons stand for celebration; and don't forget that entire language of symbols on your phone, those expressive Emojis. Esri mapping products provide a plethora of symbols to use professionally on your maps.

Classification is another thing you're accustomed to. When you group things that are in some way the same, you are classifying them. "All the girls, stand together." "Anyone that rides Bus 3, gather over here." Classifying becomes easy in a GIS because of the inherent features of a GIS system.

Symbolization in GIS is a method of assigning different sizes, colors, and shapes to features. For instance, capital cities are often represented by stars, and danger areas are usually shown in red.

Classification or styling allows GIS users to display their data by any variable that is attached to points, lines, or polygons. The variables attached are called "attributes." The following exercise gives the learner an opportunity to explore several different types of symbolization and classification/styling.

Lesson 1-3: Symbolization and classification
Nepal Earthquake Epicenters

Find the Nepal earthquake epicenters map for this lesson in the first chapter of the thearcgisbook.com. Under the ArcGIS information items heading, locate it beneath the subhead Layers.

This is a map of epicenters of the earthquakes that occurred in and around Nepal. The year of the earthquake, its epicenter, and magnitude can be viewed by clicking the points on the map. The points are also symbolized by the magnitude of the earthquake. The district divisions can be seen on the map as outlines. Now you will see how the data can be displayed differently by changing the symbolization of the map, which will allow the viewer to visualize and observe even more information.

Build skills in these areas

▶ Opening and modifying an existing online map

▶ Changing transparency

▶ Changing style to unique values

▶ Changing style to Counts and Amounts

▶ Changing size and color of symbols

What you need

▶ Account not required

▶ Estimated time: 30 minutes

Nepal Earthquake Epicenters

Scenario

The United Nations Disaster Assessment and Coordination (UNDAC) team needs an emergency response system map to respond to the Nepal earthquake. They have seen the original Nepal Earthquake Epicenter Locations map and are impressed. However, for their immediate need, they have asked that the map be altered to show the following:

- The epicenters of the 2015 earthquakes must be seen at all scales.
- All of the 75 districts should be shown by population.
- All recorded earthquakes with a magnitude of 5 and above should be shown. On the Richter scale, earthquakes above 5 can be felt by everyone and can cause slight damage to all buildings.

1. Click the Nepal earthquake epicenters map.
2. In the upper right corner, click Modify Map.

3. Click Show Contents of Map under Details

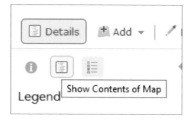

As you zoom in and out of the map you can see that the NepalEarthquake 2015 layer disappears and appears according to the scale of the map. The UNDAC wants this layer visible at all times.

4. Click the three dots at the end of the NE 0425 layer.
5. Go to Set Visibility Range.
6. Move the slider to World. This makes this layer visible at all scales.

7. Again, click Contents under Details.

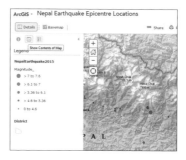

8. Zoom out to see the entire country of Nepal with the outlines of districts shown. The 75 districts are shown, but only their outlines are visible. All of the values of the districts look the same, which means they are classified on the map as location and only shows you the distribution of the data.

When you classify or style data, you have many options. The Change Style menu is your gateway to changing the look of your data.

9. Click Change Style under the District layer.

10. The individual districts can be seen more distinctly if you choose DISTRICT as the attribute. A single symbol will give you a unique symbolization by the district name.

11. All of the districts are not displayed in a unique color on the map. Click Options.
    a. Use the top slider to scroll down until you see the double arrow pointing up. Click the double arrow to Move all values out.
    b. Click Ok.
    c. Click Done. All the districts are now displayed on the map and in the Contents pane as unique values. It is now much easier to see the unique district.

You have displayed the districts by a unique value, NAME; however, the UNDAC wants the districts displayed by population. Seeing the districts displayed in a choropleth map by population would provide the responders with information about districts that would need the most resources during an earthquake. Numeric data can be displayed with counts and colors that display the features on the map as a color gradient.

12. Click District and click Change Style.
    • Select POP_91.
    • Click Done.
13. Uncollapse the District layer to see the legend.
14. Write a brief explanation of how the legend helps you understand the map.
    a. What does the legend show about the population?

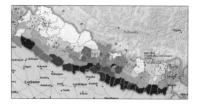

Your last task for the UNDAC is to show only the earthquakes with a magnitude of 5 and above. You want only the values of 5 and above to be shown on the map.

15. Click NepalEarthquake2015 and go to Change Style.
16. Click the Counts and Amounts (Size) options.
    • Scroll down and change the classes to 2.
    • Move the slider to 5.

This shows values 0-5 in one class and 5 and above in the Other class.

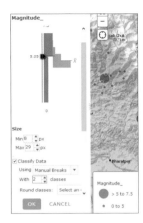

17. Click Legend.
    You might have to expand the style pane to see legend.
18. Click 0 to 5.
19. Click Fill and choose No Color.
20. Click Outline and choose No Color.

21. Click OK.

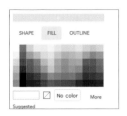

This leaves only the earthquakes with a magnitude of above 5 shown on the map.

22. Click 5 to 7.5.
23. Click Symbol, change the size of the symbol to 30, and choose a distinct symbol.

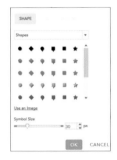

24. Click OK.
    a. Where on the map are the earthquakes with a high magnitude in relationship to a district with a high population?
    b. Turn off the District layer and observe the basemap layer. What would make rescue efforts difficult in the northern districts?

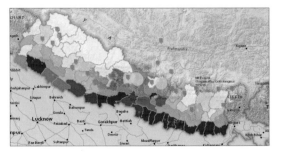

Filtering and querying
Earthquakes of the World

Again, terminology can seem more threatening than it needs to be. Filtering and querying means asking a question and, at an advanced level, asking a couple of questions strung together. A simple question would be "Where can I eat lunch?" A complex question would be "Where can I eat lunch and shop for clothes?"

A GIS system allows you to do further analysis with the data after it is mapped. In this way, GIS is more than a map. The most basic analysis in a GIS allows the user to ask questions of the data. Asking questions of the data can involve asking questions using the attributes or asking questions using the location. Questions can be constructed so they return a set of spatial results that extract meaning from your data.

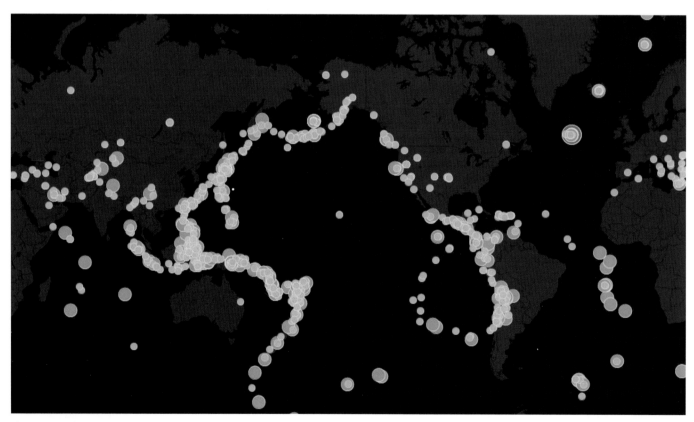

Earthquakes of the World

Lesson 1-4

Filter and query to understand earthquakes

For this lesson, find this simple live map, showing earthquakes that happened over the previous 60 days, in the first chapter of the thearcgisbook.com. Locate it under the heading "It all begins with a map."

On the map, earthquakes over the last 60 days are symbolized by magnitude and depth. Here are some basic facts about earthquakes:

- Earthquakes between 5 and 7 are the ones that occur most frequently and cause damage.
- Earthquakes that are 7 and above are extremely damaging and much rarer.
- Earthquakes deeper than 300 kilometers are usually associated with convergent boundaries.

Build skills in these areas

- ▶ Opening and modifying an existing online map
- ▶ Changing transparency
- ▶ Changing Style to unique values
- ▶ Changing Style to Counts and Amounts
- ▶ Changing the size and color of symbols
- ▶ Filtering data by attributes

What you need

- ▶ Account not required
- ▶ Estimated time: 20 minutes

Richter scale of earthquake magnitude			
Magnitude level	Category	Effects	Earthquakes per year
Less than 1.0 to 2.9	Micro	Generally not felt; recorded on local instruments.	More than 100,000
3.0–3.9	Minor	Felt by many people; no damage.	12,000–100,000
4.0–4.9	Light	Felt by all; minor breakage of objects.	2,000–12,000
5.0–5.9	Moderate	Some damage to weak structures.	200–2,000
5.0–6.9	Strong	Moderate damage in populated areas.	20–200
7.0–7.9	Major	Serious damage over large areas; loss of life.	3–20
8.0 and higher	Great	Severe destruction and loss of life over large areas.	Fewer than 3

Scenario

The United States Geological Survey (USGS) wants to use its live-feed earthquake data to promote understanding of earthquakes and their magnitude and depth. The USGS has asked you as a representative of the American Association of Geographers to write a lesson using this live-feed earthquake data, combined with the querying/filtering capabilities of the ArcGIS Online software.

Note: All answers will vary because of the last 60-day time frame of the data.

1. Click this simple live map, showing earthquakes that happened over the previous 60 days.
2. Click Show Map Contents under Details.
3. Click Earthquakes and Points to uncollapse both layers.

A multiple expression filter can be created to show only earthquakes that have a magnitude of between 5 and 7. If you have more than one expression, choose to display features in the layer that match. ALL requires that each of the criteria specified must be true.

4. Click Filter under the Points layer.

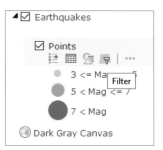

5. Choose Magnitude field from the drop-down menu.
6. Choose greater than from the Operator menu.
7. Click Unique.
8. Choose 5 from the pull-down menu.
9. Click Apply Filter.
   You now only see the spatial display of the earthquakes that meet your criteria.

10. Click Filter.
11. Click Remove Filter.
    You can ask questions that have more than one criterion. For example, you could ask to see all the earthquakes that have a magnitude of greater than 5 and a depth of more than 300 kilometers. If you have more than one

expression, choose to display features in the layer that match All or Any of your expressions. All requires the criteria you have specified must be true. Any means that only one of your expressions must be true for the features to display.

12. Click Edit.
    a. Click Magnitude is greater than 5.
    b. Add another expression.
    c. Depth (kilometers) is greater than 300.
    d. Click Value.
    e. Click All.

The map shows earthquakes that have both a magnitude greater than 5 and a depth (kilometers) greater than 300.

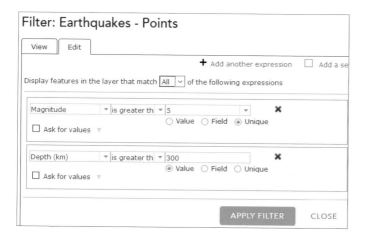

In the following lessons, you will enrich a layer to your map, add fields, and calculate values. You will also clarify the legend and publish the lesson as an app showing the rate of population change by region.

Lessons: Account required

Many of the lessons in this book do not require an ArcGIS Online organization account, but most do require saving. While users can save and share maps with either a free public account or with an organizational account, you will need an organizational account for quite a few of the lessons. If you do not have an organizational account, follow the instructions below to obtain one.

Connect with and deploy the ArcGIS platform

If you're an existing user and already have an ArcGIS subscription (with Publisher privileges), you're good to go. If you don't have these three things, continue reading.

Get a Learn ArcGIS organization membership

Some lessons in this book are carried out on the ArcGIS platform (in the cloud), and require membership (with Publisher privileges) in an ArcGIS organization. The Learn ArcGIS organization can be accessed at http://learngis. maps.arcgis.com/home/. It is available for students and others just getting started with ArcGIS. With your membership, you can immediately begin to use maps, explore data resources, and publish geographic information to the web. Go to the Learn ArcGIS organization and click the Sign up now link to activate a 60-day membership.

Getting a Learn account is the quickest and easiest way to experience web GIS at ArcGIS Online.

Using a public account

An alternative way for students, individuals, and educators who prefer not to start a 60-day trial subscription is to sign up for a Public Account at https://www.arcgis.com/home/createaccount.html. With this you will still be able to do the majority of the lessons in the book—all the "account not required" ones. The latter few that require you sign in to an organizational account appear last in each chapter, flagged by a purple vertical bar at the start of those sections.

ConnectED

In response to President Barack Obama's call to help strengthen STEM education through the ConnectED Initiative, Esri President Jack Dangermond provided a grant to make the ArcGIS system available for free to the more than 100,000 elementary, middle, and high schools in the United States, including public, private, and home schools.

ConnectED is a US government education program developed to prepare K-12 students for digital learning opportunities and future employment. This education initiative sets four goals to establish digital learning in all K-12 schools in the United States during the next few years. These goals include providing high-speed connectivity to the Internet; access to affordable mobile devices to facilitate digital learning anytime, anywhere; high-quality software that provides multiple learning opportunities for students; and relevant teacher training to support this effort.

www.esri.com/ConnectED

Lesson 1-5: Build and publish a web app
US Population Change 2000 to 2010

The U.S. Census Bureau is preparing a report on changing population trends between 2000 and 2010. They have hired your Data Visualization Company to produce a map showing rate of change by state and county during this time period. Your company has chosen to use ArcGIS Online to produce the required product. The Census Bureau has given you data for the population of 2000 and 2010 for both states and counties. They have asked for a map symbolized to distinguish areas of population growth from areas of population decline. They want to be able to have a story map web app on their website that the general public can view.

Scenario
In the following lessons you will enrich a layer to your map, add fields, and calculate values. You will also clarify the legend and publish the lesson as an app showing the rate of population change by region.

Build skills in these areas
▸ Opening a map
▸ Enriching layers
▸ Adding a field
▸ Calculating values
▸ Symbolizing the data
▸ Publishing a map as a web app

What you need
▸ User, Publisher, or Administrator role in an ArcGIS organization
▸ Estimated time: 2 hours 30 minutes

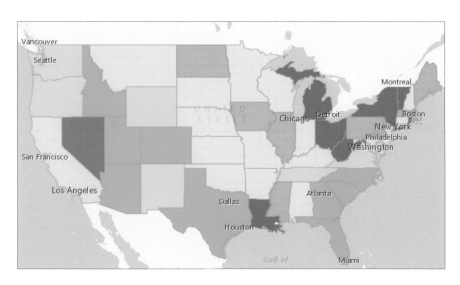

Open the map

1. Sign in to your ArcGIS organizational account.
2. Search for the Population Change ArcGIS Online group.

3. Uncheck Only search in Participants and Resources (or Only search in Learn ArcGIS).

4. Click the Population Change group.

5. Click the thumbnail of the *U.S. Population Change 2000 to 2010* map.
6. Click the Show Map Contents Button under Details.

   The map opens showing the Topographic basemap, state, and county boundaries. States and counties are political boundaries.

Change basemap

A basemap provides a background of geographic context for your map and, in this instance, the Light Gray Canvas Map shows the population data better.

1. Click Basemap on the top menu and change the basemap to Light Gray Canvas.

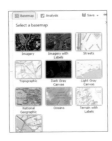

Save the map

1. On the top of the page, click Save and choose Save As.
2. In the Save Map window, type **Answers: U.S. Population Change 2000 to 2010.**
3. Type tags and a brief description of the map's content.

Show table and examine attributes

To see information about features in a layer, you can display an interactive table at the bottom of the map.

1. Show Table for States.

2. Examine the table. Notice the table shows only basic information.

3. Close the table by clicking the X in the upper right corner.

Enrich data for states

Data enrichment produces an enriched layer that retrieves information about the people, places, and businesses in a specific area. Detailed demographic data is returned for your chosen area. You are interested in a layer that shows the total population of 2000 and 2010.

1. Click States and Perform Analysis.
   The analysis icon can be activated by either clicking the analysis icon under State or by clicking Analysis in the top ribbon.

2. Click Data Enrichment and Enrich Layer.

3. Click Enrich Layer to activate the Enrich layer pane. States is the chosen layer to enrich with new data.
4. Click Select Variable to open Data Browser and browse for variables.

a. Be sure the United States is chosen in the upper right corner.
b. Click the arrow to go to the second page.
c. Click Population.
d. Click the Arrow to go to the next page.
e. Click Population Totals.
f. Click 2000 data in 2010 Geography (U.S. Census).
g. Check 2000 Total Population (U.S. Census).
h. Click 2010 Population (U.S. Census).
i. Check 2010 total Population (U.S. Census).

5. Click Apply.

6. Give Result layer a unique name.
7. Uncheck Use current map extent.
8. Click Run Analysis.

Add field and calculate: states
1. Click Show Table on your new layer. The interactive table appears at the bottom of your map.

Notice that the table now shows 2000 Total Population and 2010 Total Population. It also shows the FIPS number, which is the Federal Information Processing Standard number developed by the United States federal government for use in computer systems.

The information in the table that you need to calculate the annual rate of change from 2000 to 2010: is State_Abbr, the 2000 Total Population, and the 2010 Total Population.

2. Click Table Options in the upper right corner of the table and choose Show/Hide Columns.

3. Uncheck all the fields except State_Abbr, 2000 Total Population and 2010 Total Population.

Your next step is to add a new field to the table to store the calculation that you are going to make.

4. Click Table Option in the right corner of the table and choose Add Field.

5. Add the following parameters to the Add Field menu:
   - Name = **rate_change.**
   - Alias = **Annual rate of change from 2000 to 2010.**
   - Type = Double.
   - Click Add New Field.

You can now see the new field added to the table.

6. Click the column you have just created (Annual rate of change from 2000 to 2010) and choose Calculate.
   This opens the Expression Builder dialog box.

You are trying to find the average rate of change per year from 2000 to 2010. If you subtract the population of 2000 from the population of 2010 and divide by the population of 2000, you will have the rate of change for 10 years; if you divide that number by 10, you will have the annual rate of change; and if you multiply that by 100, you will have a percent. The formula is shown below.

7. Click the Annual rate of change from 2000 to 2010 field and click Calculate. Type or copy the following formula in the Expression Builder.

((TOTPOP10 - TOTPOP00) / TOTPOP00) / 10 * 100

8. Click Calculate.

   When you click Calculate, it populates the rows with the annual rate of change for each state.

9. Close the table by clicking the X in the upper right corner.

Symbolize and adjust legend: states

You want to distinguish your features based on the color gradient provided by the field you just calculated. The color gradient you should choose is Counts and Amounts (Color).

1. Click Enriched States and click Change Style.
2. In the Choose an attribute to show window, choose Annual rate of change 2000 to 2010.
3. Choose Counts and Amounts (Color).
4. Click Options.

5. Click Classify Data and choose Natural Breaks.
6. Choose 6 classes.
7. Click Symbols and choose Red to Green ramp.

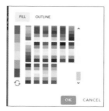

8. Click Legend.
9. For Round classes choose 0.1.

10. Click Legend and type percent symbols in the legend entries. You might have to enlarge the style pane to see legend.
11. Click OK.
12. Click Done.
13. Click States and the legend will show.
    a. Write a description of the spatial distribution of the United States by state population from 2000 to 2010.

Enrich data by Counties for States

For this exercise three states have been chosen, Virginia, Nebraska, and Arizona.

Virginia

1. Click Counties and Filter.
2. For the expression, choose:
   a. STATE_NAME
   b. Is
   c. Click Unique
   d. Scroll to Virginia
3. Click Apply Filter.

   Only Virginia counties are shown on the map.

4. Using previous knowledge enrich counties for pop2000 and pop2010.
5. Name the file.
6. Run Analysis.
7. Hide fields not needed. Keep checked Name, 2000 Total Population and 2010 Total Population.
8. Add field rate_change.

9. Use the following expression when you calculate in the Expression Builder.
((TOTPOP10 - TOTPOP00) / TOTPOP00) / 10 * 100

10. Symbolize and adjust legend.
11. Save the map to use in your web app.

Nebraska

12. Remove the filter for Virginia.
13. Filter for Nebraska.
14. Repeat steps 4 through 11.

Arizona

15. Remove the filter for Virginia.
16. Filter for Nebraska.
17. Repeat steps 4 through 11.

Create web app

You can create a web app from your map using a configurable app template. Your client has asked that the population rate change map you have built be displayed as a web app. Your client has asked you to use the configurable Story Map Series Web App.

1. Click Share.
2. Click Create a web app.
3. Select Build a Story Map.
4. Select Story Map Series.

5. Click Create app.
6. Specify a title, tags, and summary for the new web app.
7. Click Done.
8. Select Tabbed on the Welcome to Map Series Builder.
9. Click Start.
10. Type **Rate of Change Population 2000 to 2010** as the title for your Tabbed Map Series.

11. Click the arrow.
12. Add County Change for the Add tab.
13. Type **U.S. Population Change 2000 to 2010** for your map.
14. Check Legend.
15. Click Add.

16. Write an analysis of the map in the text box.

17. Add the map VA.
18. Add the map NE.
19. Add the map AZ.
20. Click Save.

21. Click Share on the top of the page.
    The Organization tab is highlighted.
22. Click View live.

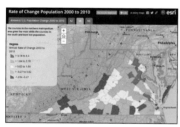

Additional analysis

The enrichment tool gives you access to a vast amount of data. For an additional learning activity, choose a variable to study and, using the above exercise as a guide, repeat the process for the chosen variable. For example, you may be interested in calculating the change in density of the population over age 65 in the last ten years. You can do this analysis either by state or at the county level.

Steps for this exercise:
1. Enrich population over 65 in 2000.
2. Enrich population over 65 in 2010.
3. Calculate population density: (2000 people over 65/area).
4. Calculate population density: (2010 people over 65/area).
    ((Density 2010-Density 2000)/Density 2000 * 100)/10.

Lesson 1-6
Use live feeds to observe the earth

Your ArcGIS Online organization account allows you access to a collection of earth observation layers that describe current conditions. These live feeds include the following:

- Active Hurricanes: Describes the path and forecast path of tropical activity from the NOAA National Hurricane Center and Joint Typhoon Warning Center.
- Recent Hurricanes: Hurricane tracks and positions, providing information on recent storms.
- Current Wind and Weather Conditions: Provides hourly data, including temperature, dew point, wind speed and direction, precipitation, cloud cover and heights, visibility, and barometric pressure.
- MODIS Thermal Activity: Presents detectable thermal activity from MODIS satellites for the last 24 hours.
- USA Stream Gauges: Provides readings of stream gauges around the United States, which depict the current water level in the measured areas.
- USA Short-Term Weather Warnings: Presents continuously updated weather warnings for the United States based on data from the NOAA National Weather Service.
- USA Wildfire Activity: Presents recent wildfire activity for the United States, featuring data from USGS GeoMac.

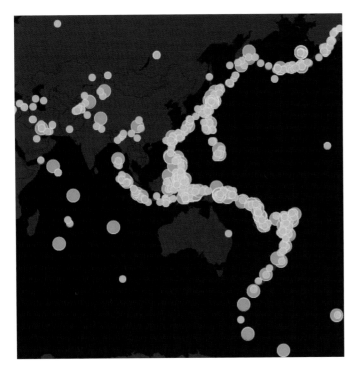

Scenario
As a GIS user you will learn how to add live-feed data.

Build skills in these areas
- Opening a map
- Adding live-feed data
- What you need
- User, Publisher, or Administrator role in an ArcGIS organization
- Estimated time: 30 minutes

What you need
- User, Publisher, or Adminstrator role in an ArcGIS Organizational Account
- Estimated time: 30 minutes

Open the map

1. Sign in to your ArcGIS organizational account.
2. Click Map on the top of the page.
3. Click the Add pull-down tab and choose Search for Layers.

4. In the Search for Layers menu, use the following parameters:

- Find: live feeds earthquakes.
- In: ArcGIS Online.
- Uncheck Within map area.

5. Select Earthquakes Live Feed from USGS.
6. Click Add.
7. Click Done Adding Layers.
   The map opens showing the Topographic basemap. A basemap provides a background of geographic context for your map, and in this case the Dark Gray Canvas Map shows the earthquake data.
8. Click Basemap on the top menu and change the basemap to Dark Gray Canvas.

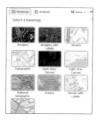

9. Refer to the lesson Earthquakes in chapter 1 for more analysis directions for this live feed.

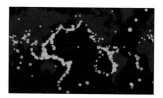

10. Remove the live-feeds earthquakes and add live-feeds weather.
11. Refer to Analyzing Real-Time Weather and Maps at https://blogs.esri.com/esri/gisedcom/2015/12/18/analyzing-real-time-weather-and-maps/ for further analysis.

12. Remove the live-feeds weather and add stream gauges.

More online lessons

At the end of chapter 1 in the online version of *The ArcGIS Book* (http://learn.arcgis.com/en/arcgis-book/chapter1/), you'll find two Learn ArcGIS Lessons. In the first lesson you will create a map of lava flow risk and in the second lesson you will analyze emergency shelters.

The ArcGIS Book, chapter 1
Questions for reading comprehension, reflection, and discussion

Teachers can use the items in this section as an assignment, an introduction, or an assessment, tailored to the sophistication of learners. Some learners can read all the sections at one time, while others are more comfortable with small segments. The questions and tasks are designed to stimulate thought and discussion.

1. Geography applied

   Thought Leader: Jack Dangermond: Web GIS is an incredible new pattern for applying geography

   a. Write an explanation of web GIS.

2. Web GIS is collaborative

   All GIS data fits onto the earth's surface:

   a. What does "Geography is the organizing key" mean?

   b. Explain what the term "georeferenced" means.

   c. This has been a fundamental concept in mapmaking for thousands of years. How has web GIS changed and expanded our use of georeferenced data?

3. The expansive reach of web GIS

   GIS is evolving: The new ArcGIS is a web GIS

   a. "ArcGIS has become a web GIS platform" is a phrase used in this section. Does this phrase raise any questions for you?

   b. What is a basemap?

4. A condensed and specific chart

   a. How is web GIS different from traditional desktop GIS? Read the material in Maps, the Web, and You and complete the following chart.

	Traditional (desktop) GIS	Web GIS
Data storage		
Data creation		
Interacting with maps and data		
Sharing and Collaboration		

5. ArcGIS information items

    a. What is the difference between a map and a scene?

    b. Investigate the maps. Pick one map and write about the information it portrays.

    c. Investigate the scenes. Pick one scene and write about the information it portrays.

    d. List three different types of layers.

Additional resources

Understanding the Difference between Consumer and GIS Mapping Applications
https://zweiggroup.com/GISsolutions/index-november-2012.php

Static Web maps vs dynamic Web GIS
http://www.pwmag.com/computerized-maintenance-management-system/static-web-maps-vs-dynamic-web-gis.aspx

Cartography Is For Everyone
Create maps that give voice to what you care about

Maps are introduced at an early age, with tried and true phrases –*X marks the spot of the treasure, This is where grandma lives, or Hansel and Gretel leave a trail of cookie crumbs*—all of which exemplify the basic purpose of a map: to show the relationship between places and the relationship of place itself to people and things.

It does not matter what media delivers a map. Whether they are hand drawn, printed, digital, or online, maps communicate. They deliver information to the viewer, information that is both analytical and artistic. Cartography is the art of making maps, and in ArcGIS Online, you can easily access a smart mapping tool to help you make the cartographic decisions that render your map informative. The tool suggests the best way to coordinate color and to represent and style your data. It allows novice users to gain experience with cartographic tools and learn from practice. When any new data is added to your online map, you'll see the smart mapping tool available on the interface.

This chapter includes a broad range of instructional topics starting with the investigation into the variety of ArcGIS Online basemaps and ending with a scenario-based lesson on the availability of farmers' markets. Within the topics, you'll find ways to add data to your maps, to analyze your data, and to better design your maps.

The chapter also includes questions that support reading comprehension, reflection, and a discussion of *The ArcGIS Book's* chapter 2 narratives.

Introductory activities

Video
The "New Rules of GIS" video is an entertaining spoof on how hard it used to be to make a map and how easy it is now. "Smart Mapping: A closer look" shows how new data-driven workflows and intelligent defaults make it easier to uncover the stories inside your data.

Go to video.esri.com and
Search for each video by the title.

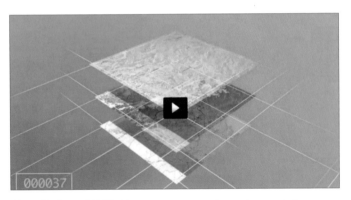

New Rules of GIS: How savvy people make and share maps

Smart Mapping: A closer look

Activity

When deciding how to organize their time around the lessons in this chapter, students and teachers might want to combine the next two lessons (the lesson on basemaps and the one on scale and resolution) as an activity.

Account not required

The first four lessons in chapter 2 cover a broad range of topics that are instructional, interpretive, and skill based. The first two lessons are short and more instructional than scenario driven. The topics are as follows:

1. An investigation of basemaps
2. An explanation of the difference between scale and resolution
3. Investigation of a map with sophisticated and creative classification
4. Adding spatial components to a map:
   - x,y (latitude and longitude)
   - addresses (geocoding)

Lesson 2-1:Investigate basemaps

Identifying the geographical context of your content

A basemap provides a background of geographical context for the content you want to display in a map. When creating a new map, you first choose which basemap to use. And you can experiment. You can change the basemap of a current map at any time by using the basemap gallery. The purpose of the basemap is to provide an appropriate background setting for your work. In order to make a decision as to what basemap to select, start by becoming familiar with all the basemap layers. This will enable you to recognize the basemap that is appropriate for your subject, your scale, and the type of data you are using.

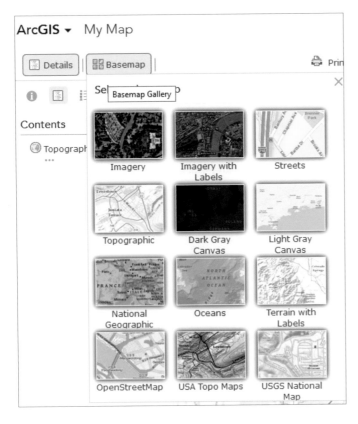

Build skills in these areas

▸ Opening and modifying an existing online map

▸ Changing basemaps

▸ Making decisions about the selection of a basemap

What you need

▸ Account not required

▸ Estimated time: 30 minutes

Scenario

Deciding which basemap to use with your data requires knowing what basemaps are available and making judgments as to which basemap is right for your data. You have been asked to talk about selecting basemaps. Before you can make your presentation, familiarize yourself with the 12 Esri basemaps available.

1. Go to ArcGIS Online.
2. On the top of the site, click Map.
3. Click Content under Details.

4. The default basemap is Topographic. Click twice and click the three dots (elipsis) under Topographic and click Description.

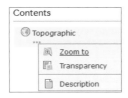

5. When you click Description, it takes you to a description of the basemap layer.

6. On the top of your browser, return to the ArcGIS—My Map tab and investigate the Topographic basemap layer.
   - Zoom to the world.
   - Zoom to the county level.
   - Zoom to a neighborhood level.

7. Click Basemap on the top of the page. This exposes icons to the 12 Esri basemaps provided as a background for your work. Become familiar with all the basemap layers.

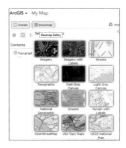

8. Pick three basemaps and write a three-word description for each.

   a. Imagery:_____
   b. Imagery with Labels: _____
   c. Streets:_____
   d. Topographic: _____
   e. Dark Gray Canvas: _____
   f. Light Gray Canvas:_____
   g. National Geographic: _____
   h. Oceans: _____
   i. Terrain with Labels: _____
   j. OpenStreetMap_____
   k. USA Topo Maps: _____
   l. USGS National Map: _____

9. Basemaps are described as multiscale.
   a. What would basemaps look like if they were not multiscale?

Lesson 2-2: Scale and resolution
Mastering the difference between them

Maps have always been defined as a way to communicate a graphical representation of the earth's surface, so mapmakers have always dealt with scale. With the introduction of digital orthophotos, the property of resolution became equally important to consider. Resolution is a linear dimension on the ground that is represented by each pixel. If resolution is high, you will be able to see more detail in a digital photo as you zoom in; if resolution is low, you will be able to see less detail.

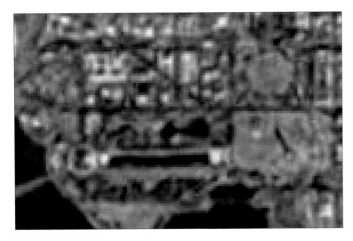

Low resolution

High resolution

Build skills in these areas

▸ Opening and modifying an existing online map

▸ Finding and interpreting a scale on a basemap

▸ Observing and defining resolution

▸ Explaining the difference between scale and resolution, verbally and in writing

What you need

▸ Account not required

▸ Estimated time: 30 minutes

Scenario

You have been asked to give a presentation to beginning teachers about the difference between map scale and resolution. The designated audience is middle school students.

1. Go to ArcGIS Online.
2. On the top of the site, click Map.
3. Click Basemap and change the basemap to Imagery.

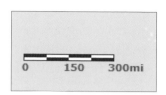

4. Investigate scale by answering the questions below:
   a. Where is the scale shown on the map?
   b. In the Find tab on the upper right, search for the following locations and zoom in as far as possible. Are there different zoom restrictions?
      • Washington, DC
      • Mount Kilimanjaro, Kilimanjaro, Tanzania
      • Moscow, Russia
      • Seattle, WA
   c. Are the images clear at the last zoom?

5. Resolution by definition is a function of the satellite or aerial imagery used. A detailed description of the imagery resolution is found at http://www.arcgis.com/home/item.html?id=10df2279f9684e4a9f6a7f08febac2a9
   a. In one sentence summarize what the resolution is after reading this description.

6. Verbalize to a small group the difference between scale and resolution. Feel free to use the maps as visualizations.

Lesson 2-3: Classification

2008 presidential election results by precinct

For use in this lesson, find the *Past as Prologue? 2008 presidential election results by precinct* map in chapter 2 of thearcgisbook.com. It is the first map under the heading What maps can do.

Maps communicate and foster understanding by providing access to data. Data is classified so information can be communicated. In GIS data is communicated in choropleth maps, and thanks to the skills of the analyzer and cartographer, you can delve deeply into all kinds of geographic phenomena visually.

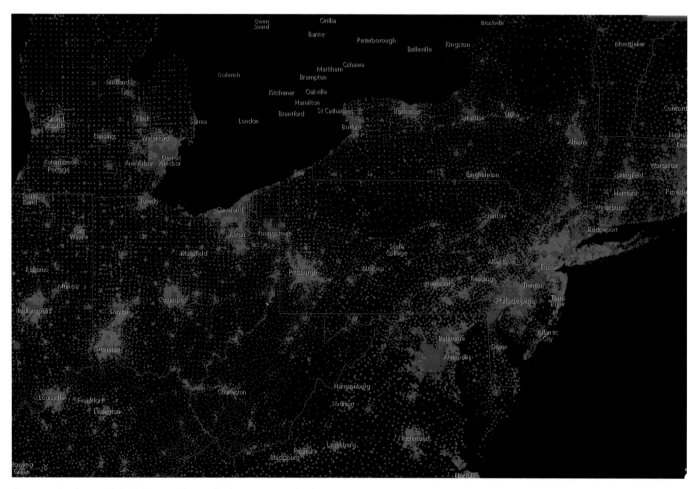

Past as Prologue? 2008 presidential election results by precinct

Scenario

The Democratic and Republican candidates for the 2016 presidential elections have decided to make use of this map's data to help them decide where to concentrate their campaign efforts. Florida, Virginia, Ohio, Iowa, Colorado, and Nevada have been declared swing states. As a political intern you have been tasked to pick a state and analyze the data to decide what geographical areas would be best suited for your candidate to campaign. You may pick either the Republican or Democratic party.

Build skills in these areas

▸ Opening and modifying an existing online map

▸ Interpreting a map with sophisticated classification

▸ Using a map to make predictions

What you need

▸ Account not required

▸ Estimated time: 30 minutes

1. Click Past as prologue? 2008 presidential election results by precinct to open the story map.
2. Read the sidebar information.
3. List the layers of information shown on the map.
4. Pick a state and zoom to it.
5. Using the map data, write your analysis including the following:
   • Geographic locations
   • Voting patterns by precincts
   • A reference to ethnicity
   • A reference to income
6. Take screen captures of the maps and include them in your written analysis.

Lesson 2-4: Adding point data
Find patterns in mountains of data

For use in this lesson, find *Maps help find patterns in mountains of data* in chapter 2 of thearcgisbook. com. It is the fourth map under the heading What maps can do.

You begin to get a strong sense of what maps can do from this exercise. The image shows a total of 58,000 airline routes on one map. The Web Mercator projection transforms the straight flight paths into curves. You are going to use this map to solve a spatial problem while also learning how GIS software deals with point data. Locational information contained in point data adds the spatial component that the map needs to locate the information. Point data can get its locational information from latitude and longitude or it can get it from a street address.

Build skills in these areas

▸ Opening and modifying an existing online map
▸ Filtering
▸ Adding x, y data
▸ Geocoding data
▸ Creating bookmarks

What you need

▸ Account not required
▸ Estimated time: 30 minutes

Maps help find patterns in mountains of data

Scenario

The International Federation of Air Traffic Controllers' Associations (IFATCA) wants to investigate flights going in and out of three of the busiest places in the world: London, New York, and Atlanta. They want a visual representation of the flight routes to each of the airports within the three cities.

Open the map, add and filter data

1. Click *Maps help find patterns in mountains of data.*
2. In the upper right corner, click Modify Map.

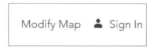

3. In the top menu, under Details, click Content. You should see two layers in the Content pane: World Dark Gray Base and flight routes.

   You have been asked to map three individual cities that have the largest air traffic: London, New York, and Atlanta. You need to add a layer to obtain that information.

4. Click Add on the Top menu and choose Search for layers.

5. When the Search for Layers menu appears, use the following parameters:
   - In the Find box, type
     boundaries,places,cities,towns,world,people
   - In: ArcGIS Online
   - Uncheck Within map area

6. Add World Cities by esri.
7. Click Done Adding Layers.

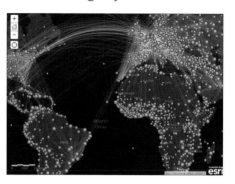

You now want to narrow down the cities to the three individual cities: London, New York, and Atlanta. You do this by applying a filter, which presents a focused view of the feature layer World Cities.

8. Point to World Cities and click Filter.

9. Create your definition expression. Be sure to select Any.

   - CITY_NAME is London.
   - Be sure to click Unique.
   - + Add another expression.
   - CITY_NAME is New York
   - + Add another expression.
   - CITY_NAME is Atlanta.
   - Click Apply Filter.

Once again at the top of the Filter Menu be sure you change the setting from All to Any.

Locate and map airports

Now that the three cities are isolated, it is time to locate and map the individual airports.

In order to add the location of the individual airports, you will deal with two types of point data.

Type 1: Add the first set of airports to the City of London by using latitude and longitude (x,y) data.

Type 2: Add the second two cities (New York and Atlanta) with address information.

Add this feature data from a delimited text file (.txt) or a comma-separated values text file (.csv) that includes the locational information (latitude and longitude or address) along with any other attributes.

What follows is an example of a delimited text file for the airports for the City of London with latitude and longitude data. Notice how the headers are separated by commas with no spaces. That is why it is called a comma-

delimited text file. The name of the airport is the only attribute attached to the locational information. Notice also that the latitude and longitude are in decimal degrees. These must be exact and on one line.

Long,lat,name
-0.45361,51.47196,Heathrow
-0.17899,51.15518,Gatwick
-0.37557,51.8798,Luton
0.23985,51.88516,Stansted
0.70171,51.56612,Southend

10. Type the above lines exactly as shown into a simple text application. Notepad is a great application to use for this. Be sure to copy the top line: long,lat,name.

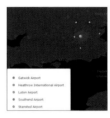

11. Save the file on your computer and name it **airports_x_y.**
12. Drag the airport_x_y file to your map.
    This is the power of GIS. You have taken a table of data and mapped it. If you click on any of the points, you can see the locational information as well as the name of the airport.

Change symbology

13. The airports are displayed by unique symbols with the attribute as the name. The symbols are too small for the map. Under Types (Unique symbols), click Options.

14. Click the color ramp to activate the change symbol menu.
15. Move the Symbol Size to 15 and choose a shape.

16. Click OK.
17. Click Done.

Create bookmarks

To get quickly from airport to airport on your map, as the map author, you need to create bookmarks. Bookmarks are based on the current location and scale of the map. When others click a bookmark, the map zooms to that location.

18. Zoom to the location and scale that you want to look at London and its airports. You might want to change the basemap.
19. On the top of the site, click Bookmarks.
20. Click Add Bookmark.
21. Name the bookmark **London.**

22. Click the X in the upper right corner to close the bookmark.

Add the second two cities (New York and Atlanta) with address information. Geocoding is the proper term for converting an address to an x,y coordinate. By default, ArcGIS Online uses the World Geocoding Esri service.

Below is what you need for a comma-delimited text file for New York and Atlanta.

Name,address,city,zipcode
Laguardia,LaGuardiaRd,Flushing,NewYork,11371
JohnF.Kennedy,Jamaica,NewYork,11430
Hartsfield-Jackson,6000N.TerminalPkwy,Atlanta,GA30320

23. Type the above lines exactly as shown into a simple text application (like Notepad).
24. Save the file on your computer and name it **airports_addresses**.
25. Drag the airport_addresses file to your map and add layer.
26. Change the size of the symbols.
27. Zoom to New York and create a bookmark. Name the bookmark **New York**.
28. Zoom to Atlanta and create a bookmark. Name the bookmark **Atlanta**.

By completing this map you have given the IFATCA a visual representation of the flight routes to London, New York, and Atlanta. In order to do that, you used the following GIS skills:
- Searching for ArcGIS Online layers
- Filtering layers
- Mapping table data with an x,y coordinate
- Mapping table with an address
- Creating bookmarks

Lesson 2-5: Derive accessibility
Can you stroll to the farmers' market?

Farmers' markets are one way people have access to healthy food. The *Farmers Markets* map shows the location of farmers' markets. The map is designed for learners to use to answer the question: how many Americans live within a reasonable walk or drive to a farmers' market. For analysis purposes, a 10-minute drive time and a 1-mile walk time will be used to determine what is reasonable.

Scenario

Many county chambers of commerce have been asked to sponsor farmers' markets. In this lesson, you will provide maps for county officials showing the location of markets within their counties and reasonable drive and walk times from the markets. You have the freedom to choose from three different counties. The county may represent different geographic areas in your state or you might choose to compare counties from other states. Once you choose a county, derive drive times and walk times of markets within the county. Last, you will clarify the legend and publish the lesson as an app showing the availability of markets in the three different counties. It is your plan to present your map and analysis to the individual chambers of commerce.

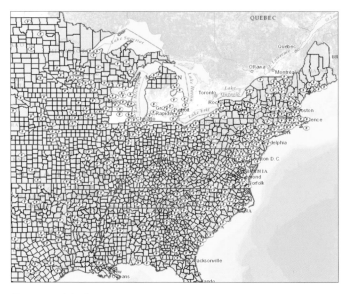

Farmers Markets.

Build skills in these areas

▸ Opening a map

▸ Filtering a layer by county

▸ Deriving drive times

▸ Deriving walk times

▸ Symbolizing the data

▸ Publishing as a web app that compares three different counties

What you need

▸ User, Publisher, or Administrator role in an ArcGIS organization

▸ Estimated time: 1 hour 30 minutes

Open the map

1. Sign in to your ArcGIS organization account.
2. Search for the Farmers Markets ArcGIS Online group.

3. Uncheck Only search in Participants and Resources.

4. Click the Farmers Markets group.

5. Click the thumbnail of the Farmers Markets map.

   The map opens showing the Topographic basemap, state, and county boundaries. States and counties are political boundaries.

Save the map

1. On the top of the site, click Save and choose Save As.
2. In the Save Map window, type **Answers: Farmers Markets.**
3. Type tags and a brief description of the map's content.
4. Click Save Map.

Select and filter a specific county (Fairfax, VA)

A filter presents a focused view of a feature layer in a map. In this particular case you only want to view and analyze the farmers' markets that are in Fairfax, VA. Fairfax County is the most populous county in the Commonwealth of Virginia and contains 13.6 percent of Virginia's population.

1. Click show content of map under details. Point to Farmers Markets and click filter.

2. Create your definition expression.
   a. County is Fairfax (click Unique).
   b. Add another expression.
   c. State is Virginia.
   d. Select All.

3. Click Apply Filter.
4. You should now see the farmers' markets in Fairfax County.

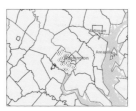

5. Click to save map. Showing the table for Farmers Markets, you will see there are 19 of them within Fairfax County.

6. You are now ready to perform analysis and derive the 10-minute drive time and the 1-mile walk distance from each of the farmers' markets. Click the Perform Analysis Tab.

7. Click Use Proximity.
8. Click Create Drive-Time Area and use the following parameters:
   - Choose point layer to calculate drive-time areas around = Farmers Markets
   - Driving Time = 10 Minutes
   - Areas from different points = Split
   - Result layer name = 10min_drive
9. Click Run Analysis.

10. Click Perform Analysis on the Farmers Markets layer.
11. Click Use Proximity.
12. Click Create Drive-Time Areas and use the following parameters:
    - Choose point layer to calculate drive-time areas around = Farmers Markets
    - Walking Distance = 1 Mile
    - Areas from different points = Split
    - Result layer name = 1mile_walk
13. Click Run Analysis.
14. Right-click filter under Farmers Markets.
15. Click Remove Filter and all the farmers' markets are visible.
16. Save as Fairfax, VA, Farmers Markets.
17. Share with Everyone.

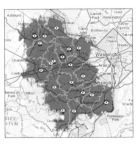

Select and filter a specific county (Fulton, GA)
Fulton, GA, is the county home of Atlanta, which is the capital and the most populous city in the state of Georgia.

1. Click the feature layer Farmers Market and click Filter.
2. The definition expression should be the following:
   - County is Fulton.
   - Add another expression.
   - State is Georgia.
3. Repeat steps 3 through 17 as you did for Fairfax County. There should be 10 farmers' markets. There are two farmers' markets very close together at the northernmost point of the county.

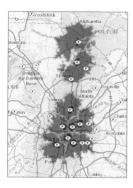

4. Save as Fulton, GA, Farmers Markets.

Select and filter a specific county (Dallas, TX)
Dallas, TX, is the second most populous county in Texas and the third-largest city in Texas.

1. Click the feature layer Farmers Market and click Filter.
2. The definition expression should be the following:
   - County is Dallas.
   - Add another expression.
   - State is Texas.
3. Repeat steps 3 through 17 as you did for Fairfax County. There should be 6 farmers' markets. There are 2 farmers' markets very close together at the northernmost point of the county.

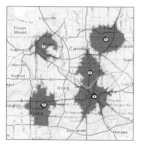

4. Save as Dallas, TX, Farmers Markets.

Create web app

You can create a web app from your map using a configurable app template. Your client has asked that the population rate change map you have built be displayed as a web app. Your client has asked you to use the configurable Story Map Series web app.

1. Click Save to be sure your work has been saved.
2. Click Share. Share with your Individual Organization or with Everyone.
3. Click Create a Web App.
4. Click Build a Story Map.
5. Click Story Map Series.
6. Click Create App.
7. Specify a title, tags, and a summary for the new web app.
8. Click Done.
9. Choose Tabbed on the Welcome to Map Series Builder.
10. Click Start.
11. Click Settings on the top of the page.
12. Click Map Options.
13. Uncheck Synchronize map locations.
14. Click Apply.
15. Add Farmers Markets Accessibility for the Tabbed map Series
16. Click the arrow.
17. Tab title = Dallas, TX.
18. Content should have Map Checked.
19. For Map Select the map of Dallas, TX Farmers Markets.

20. Click Add.
21. Write an analysis of the map in the text box.
22. Click Add.
23. Add Fulton, GA, Farmers Markets to the tab.
24 Select a Map.
25. Choose Fulton, GA, Farmers Markets.
26. Write an analysis of the map in the text box.
27. Repeat steps 18 through 22 for Fairfax, VA.
28. Click Save in the upper right corner.
29. Click Share on the top of the page, The Organization Tab is highlighted.
30. Click View live.

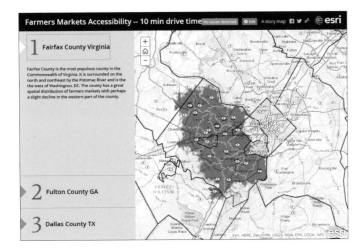

More online lessons
At the end of chapter 2 in the online version of *The ArcGIS Book* (http://learn.arcgis.com/en/arcgis-book/chapter2/), you'll find a Learn ArcGIS Lesson predicting deforestation by prohibiting a proposed road construction.

The ArcGIS Book, chapter 2
Questions for reading comprehension, reflection, and discussion

Teachers can use the items in this section as an assignment, an introduction, or an assessment, tailored to the sophistication of the learners. Some learners can read all the sections at one time, while others are more comfortable with small segments. The questions and tasks are designed to stimulate thought and discussion.

1. The online mapping revolution
   This section refers to consumer maps on smartphones and the web as being the most regularly used applications on smartphones and mobile devices.

   a. List three of these applications.

2. GIS maps engage an audience for a purpose

   a. Of the six different maps shown, which two are the most critical to human life?

3. What maps can do
   a. There are six maps, each one illustrating a different purpose. Choose a map that interests you and summarize the map in a short paragraph.

4. The role of web maps: At their heart, web maps are simple basemaps and operational layers

   a. Explain the purpose of basemaps compared to operational layers.

5. Web map properties: Continuous and multiscale

   a. Besides being scalable and fluid, name two other advantages of online maps.

6. New smart mapping workflows

   • **Thought Leader: Jim Herries: Map design is about drawing your audience into the story you're telling**

   • **ArcGIS for cartographers: The art of mapmaking**

   • **Representing elevation and terrain: Using ArcGIS for Desktop**

   a. What does elevation or a terrain add to a map?

Additional Resources

Design principles for cartography, Esri blog
https://blogs.esri.com/esri/arcgis/2011/10/28/design-principles-for-cartography/

Making Maps: DIY Cartography, Resources and Ideas for Making Maps, blog
https://makingmaps.net/

Tell Your Story Using a Map
Publish maps that engage people through storytelling

Whether it's around a campfire, in a classroom, or over a cup of coffee, everyone loves a good story. Esri first launched story maps in 2012, and they've become very popular since then. Partly because of them, ArcGIS Online has progressed from being a pioneering tool for online spatial visualization and analysis to a groundbreaking tool for sharing maps, data, and the results of spatial analysis online. Story maps describe places, reveal change over time, communicate breaking news, and recount history. They have already effected change, influenced opinion, created awareness, and sounded the alarm about impending threats.

With their ability to combine text, photographs, audio, and video data on a live (interactive) map, story maps are fostering nothing short of a communication revolution. They allow anyone to see, and therefore understand, even complex information quickly, within an accurate spatial context. Story map templates enable you to create a visually engaging narrative whether or not you know GIS because almost everything you need comes with them. Story maps bring the tools and analysis capability of GIS to you, without requiring you to be skilled or knowledgeable in the GIS software functionality. Yet you can use them to begin to learn GIS in a way that is fun.

This chapter offers practice in the creation of several types of story maps through four scenario-based lessons. By following the step-by-step instructions for creating one or more story maps in each lesson, you will also be building or reinforcing your own fundamental ArcGIS Online skills. Use the questions at the end of this chapter to support your reading comprehension, reflection, and discussion of the narratives presented in the corresponding chapter 3 of *The ArcGIS Book*.

Introductory activities

Video

What makes story maps different from other web maps? Allen Carroll explains how cloud computing, data sharing, and the proliferation of mobile devices have permanently changed the landscape of map creation and combined to produce an enormous opportunity for us to tell our stories in a whole new way. Story maps meld maps, data, and multimedia content to communicate effectively. Carroll provides an overview of the variety of Esri Story Map templates and suggests potential uses for each. Use this video as a starting point to explore the range of templates and examine their applicability to different topics and purposes.

Watch a video on storytelling and information design
Go to video.esri.com and search for "Every Map Tells a Story."

Activity

Explore the Story Maps Gallery

Go to storymaps.arcgis.com to view story maps by app type or by subject. You can also use the search window to look for specific topics such as World War II, food, or disease. As you explore the maps in the gallery, think about the subjects you like to study or teach. Find three maps that you could use for further learning in your chosen areas. For each map, drill down to identify why you think it's suitable, by answering the following questions:

1. What is the name of the story map?
2. Who do you think the intended audience for this story map is?
3. What do you think the author wanted to convey to his/her audience?
4. What subject of study or discipline you would you use it in?
5. How does the message this story map communicates support your own educational objectives?

Studio 615 - A Case Story

By Datastory

The Amazonian Travels of Richard Evans Schul

By The Amazon Conservation Team

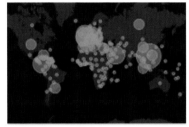

Panama Papers: Mapped

By Esri UK

The Great In-Between: A Mongol Rally Map Jou

By Cooper Thomas, Esri

Great Places in America

By American Planning Association

On the Brink: The Sixth Great Extinction

By Esri Story Maps team

Galeria De Mapas de Diario De Cuba

TransNet Keeps San Diego Moving

Development in NoMa & Union Market

Lesson 3-1: Create story maps
London's 1854 cholera epidemic twice told

The variety of Esri Story Map templates allows you to tell a story in many ways and to shape the viewer's experience. It's important to choose a template that reflects your purpose in creating the web application. This lesson will walk you through the steps of creating two types of story maps about the same historic event.

Dr. John Snow is regarded as one of the founding fathers of modern epidemiology. During a major cholera epidemic in 1854 London, he collected and mapped data on the locations (street addresses) where cholera deaths occurred. His process was laborious and slow, but ultimately very informative. His painstakingly detailed analysis led to the identification of the epidemic's source. Today, John Snow's data has been geocoded making it accessible in a geographic information system. In this lesson you will create a heat map showing the locations that experienced the highest number of cholera deaths in the epidemic in 1854 London. You will share this heat map as a web application using two different story map templates.

Scenario

As part of a pre-med curriculum, you are taking a course in the history of epidemiology. Your first assignment is to prepare a presentation about Dr. John Snow's investigation of cholera during an 1854 outbreak in London. Luckily for you, John Snow's 19th-century data has been geocoded to make it accessible in a geographic information system. You plan to include a web map in your presentation that reflects Dr. Snow's conclusion that contaminated public water pumps were responsible for the outbreak. However, you can't decide which of two web app formats is most appropriate for sharing Dr. Snow's story. The best way to make up your mind is to create both apps so you can compare them.

Build skills in these areas

▸ Opening a map

▸ Changing the style of point data to create a heat map

▸ Creating a basic story map

▸ Creating a Map Tools web application

What you need

▸ An ArcGIS Online account, (free public or organizational)

▸ Estimated time: 1 hour

Explore a map of London's 1854 cholera epidemic

1. Sign in to your ArcGIS Online account.
2. Click search.
3. Search for LearnResource.
4. Click Details under the LearnArcGIS Cholera.
5. Read the description of the map to familiarize

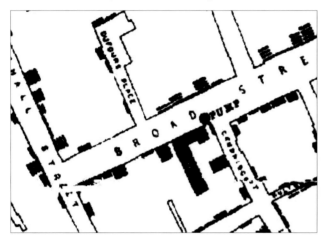

   yourself with this famous historical investigation
6. When you have finished, click the thumbnail to open the map.
7. Use Save As to save it to My Content in your ArcGIS Online account.
8. Click the Contents tab to see the map's layers.
9. Zoom in until you can see the little hash marks at points along the streets.
   Each of these marks indicates one or more cholera cases (deaths) at a particular address. As you can see, some addresses had only one or two deaths, but many addresses had more.

10. Use the Soho bookmark (click Bookmarks, and select Soho) to return to the full map of the Soho district.

The cholera cases by address are the same addresses Snow mapped, but they have been geocoded to be accessible in a GIS. Each red dot is an address and has data associated with it called attributes. One of the attributes for this point layer is Num_Cases. This is the number of cases (deaths) that Snow recorded at this address with his hand-drawn hash marks.

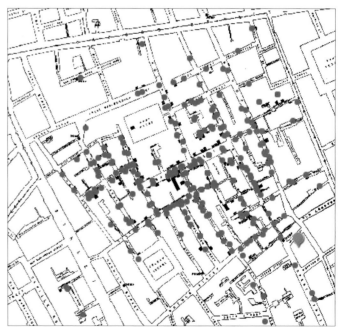

11. Click on any red dot to see its attributes in a popup window. The popup provides the address and the number of cases at that address.

Create a heat map

Snow collected data and recorded it on his map so he could determine where the greatest number of cholera deaths occurred. This was a laborious process for Dr. Snow, but today GIS provides tools that help us answer this question quickly. Geocoding the addresses and cholera data opens the door to a large number of analysis processes in GIS. We'll employ only one of these analysis processes for now. We'll analyze the location of cholera deaths by creating a heat map. Heat maps use point layer data to calculate and display relative density. The colors are most intense where the most points are concentrated together. In this case, we don't want to know where the greatest number of dots is but where the greatest number of cases is.

1. Turn on Public water pumps.
   Point to the name of the layer, Cholera cases by address. When you do this you will see a series of icons appear beneath the layer name. If you point to an icon, you will see what it does.
2. Click Change Style.

3. Now you will change the map style to a heat map.
   a. Under Choose an attribute to show, select Num_Cases. In other words, you are telling the program that you want to create a map that focuses on the number of cases at any location.

b. Under Select a drawing style, click Heat Map, and click Options.

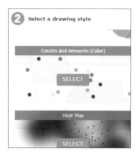

c. Move the Area of Influence slider slightly to the right to enlarge the area of highest density on the map.

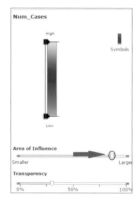

4. Click OK.
5. Click Done.
   Your map now looks like this. The yellow area indicates the part of Soho where the greatest number of cholera deaths occurred. As you move through orange, red, and purple to blue areas, the density of cholera cases drops off.

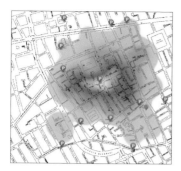

6. Click the pump symbol in the center of the yellow area to learn its name.

This is the famous Broad Street pump. Snow's painstaking analysis of the same data led him to the conclusion that the Broad Street pump was the source of the Soho epidemic. When the pump handle was removed, preventing people from using it, the epidemic came to an end. Through spatial analysis, many lives were saved and cholera was confirmed as a water-borne disease. Before you create your story map, make the following adjustment to your map.

7. Click the dots below John Snow's 1854 map of cholera cases and select Hide in Legend. This will simplify the legend in the story map you will create.

8. Save the map.

Create a basic story map

1. To begin creating a story map, click Share (above the map).
   a. Choose Everyone to make your map public.
   b. Click Create a Web App under Embed this map.

2. Click Build a Story Map.
3. Click Story Map Basic.
4. Click Create App.

5. In the web app window, fill in the following information for your map:
   a. Title: **London Cholera Epidemic 1854, [your initials] Map 1** (for example, London Cholera Epidemic 1854, LM Map 1)
   b. Summary: **Heat map of Cholera cases in the Soho District of London, 1854.**
6. Click Done.
   You now see a configurable version of your new story map.
7. Under Settings, type the title you just used when creating the story map (**London Cholera Epidemic 1854, [your initials] Map 1**).

As you can see, you can change the subtitle (it defaults to the map summary), create links from the headers, set the color theme, and choose whether to include a legend or a search tool in your map.

8. Experiment with different settings.
   a. Click Save and click View to see what your map will look like.
      You can continue to make changes to the settings. Click save to see the result of each set of changes.
9. When you are satisfied with your story map, click Close.

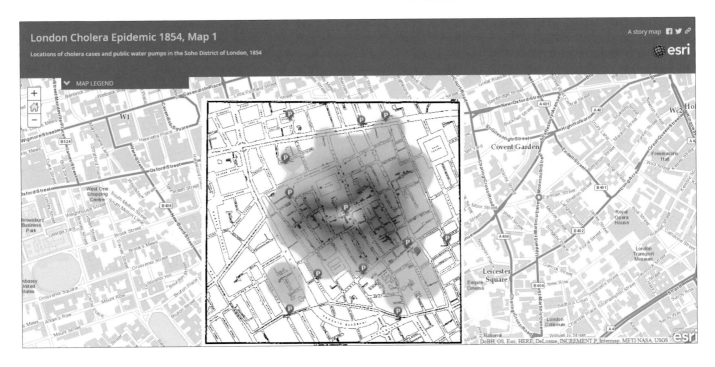

Create a map tools web app

A basic story map is one way to share the story of London's 1854 cholera epidemic. You design the map you want the viewer to see. Sharing your map as a Story Map Basic allows the viewer can see but not modify the map in any way. This time you will use the same cholera map to create an interactive story map. The viewer will be able to turn layers on and off and explore the data behind the map.

1. Sign in to your ArcGIS account again.
2. Open LearnArcGIS Cholera_LM.
3. Click Share.
4. Click Create a Web App.
5. Choose the Map Tools template (in the All section).
6. Click Create App.

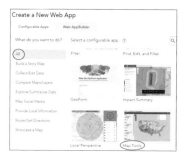

7. Provide information about your app
   a. Change the title to **London Cholera Epidemic 1854, [your initials] Map 2** for example, London Cholera Epidemic 1854, LM Map 2).
   b. Fill in the Summary as follows: **Heat map of Cholera cases in the Soho District of London, 1854.**
8. Click Done.

9. Provide map title and description.
   This time you have a greater number of options than you did in the Story Map Basic app.
   a. Type the following in the Title box: **John Snow's Investigation of Cholera, London 1854 Map 2.**
   b. Type the following in the Description box: **Heat Map of Cholera cases during the 1854 London epidemic in the city's Soho district.**

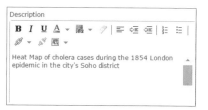

10. Configure map settings.
    a. Indicate selections for theme colors.
    b. Under options, check the following tools to boxes to add your app: Basemap Gallery, Home Button, Legend, Scalebar, and Zoom Slider.

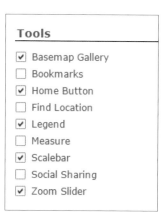

    c. Under Layer list, uncheck Include sub layers in Layer list.
    d. Leave the default Table Settings.
    e. Click search and uncheck Enable search tool.
    f. Click Print and uncheck Print Tool and Display all Layout Options.

10. Click Save.
11. Click Close.
Your new story map looks like this.

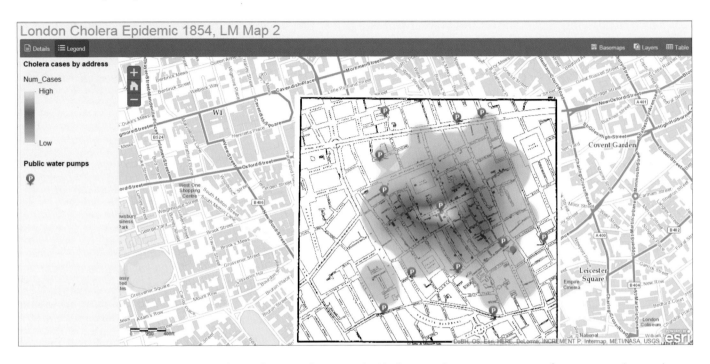

You can close the Legend by clicking the word Legend. Click Details to see more information about the map. (Such information comes from the description of the original map.)

Extended activity
- Save your cholera map again but give it a different title. Use it to experiment with alternate story map and web app templates to see how many you can find that are appropriate for telling the John Snow story.
- Use additional analysis processes (styling, filtering, buffering, and so on) to further explore the geocoded Snow data.

Resources
- *John Snow and the Broad Street Pump on the Trail of an Epidemic,* http://www.ph.ucla.edu/epi/snow/ snowcricketarticle.html
- *The Ghost Map: The Story of London's Most Terrifying Epidemic—and How it Changed Science, Cities and the Modern World* by Steven Johnson.

Lesson 3-2: Create a swipe story map
Are graduation and unemployment rates related?

In the effort to answer questions and solve real-world problems, it is frequently necessary to explore the relationship between different variables. For example, to answer a question about whether there is a relationship between diabetes and obesity, the investigator needs to compare the rates of these two conditions. Spatial data – data tied to a specific location – is essential in this. Mapping rates of diabetes and obesity reveals patterns of distribution that suggest a definite connection between the two. Beyond visual observations of patterns, exploring this data in a GIS means it is also possible to query the data and employ a range of analysis tools to further explore the relationship.

Story maps provide a number of ways to compare maps and data. The Story Map Swipe and Spyglass app enables users to interact with two web maps or layers simultaneously in a single scalable view. In this lesson you will create a swipe map to compare high school graduation rates with unemployment rates.

Scenario

How important is it for future members of the workforce to complete high school? What is the relationship between graduation rates and unemployment? A national education advocacy group focused on reducing high school dropout rates has hired your Data Visualization Company to produce a web map that can be imbedded on their website comparing high school graduation rates with unemployment rates.

Build skills in these areas

▸ Opening a map

▸ Creating a swipe story map

What you need

▸ An ArcGIS Online account (free public or organizational)

▸ Estimated time: 45 minutes

Open a map with employment and graduation rate layers

1. Sign in to your ArcGIS Online account.
2. Click search.
3. Search for LearnResource.
4. Click the thumbnail of the Graduation Rates and Unemployment map to open it.
5. Use Save as to save to your own account as **Graduation Rates and Unemployment-[your initials]**.

   The two layers in this map are the ones you will use in your swipe story map. Be sure both are turned on.

Create a swipe story map

1. Click Share (above the map).
2. Click Create a Web App under Embed this map.

3. Under Build a Story Map, select Story Map Swipe and Spyglass.

4. Click Create App.
5. Type the following in the summary box:
   Choropleth maps of US high school graduation rate, and US unemployment rate.

6. Click Done.

   In the Swipe/Spyglass Builder, notice that the Vertical bar is the default selected layout. This is the one you want.
7. Click Next.

In the next window, notice that A layer in a web map is the default selected layout. Again, this is the one you want.

8. Click Next.

   Note: the Builder has identified Unemployment rate 2010 as the layer to be swiped. This just determines which of the two layers will be on the bottom and which on the top.

9. Accept the default layout settings and click Next.

10. Select colors for your map headers and give each header a title.
Note: the Right Map will be the same as the one Builder identified as the map to be swiped.

11. Click Open the app.
12. Add a description of the swipe map in the box that says Edit me.
13. Modify the title or the subtitle (this is the map summary) if you wish to do so.
14. When you have finished, click Save.
Your completed swipe map opens like this:

If you zoom in and move the swipe bar you can observe relationships at both the regional and local scales.

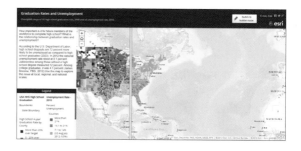

Extended activity
Create another swipe map using census and/or health data to explore possible relationships among variables, For example, race and income, infant mortality rate, and poverty.

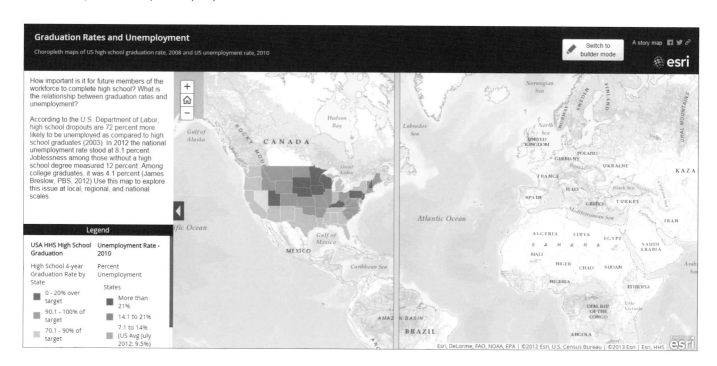

Lesson 3-3:Create a Map Series web app
World's Largest Cities 100 – 2000 CE

ArcGIS Online Story Maps provide an excellent vehicle for portraying events over a span of time. Changes in land use, the spread of disease, the growth of US railroads, and urban sprawl are but a few topics that lend themselves to such story map presentations.

The Map Series Story Map is a useful vehicle for telling a story that evolves over time. In this lesson, you will create a Tabbed Story Map Series to show the 10 largest world cities in different periods of time.

Scenario

You have been hired to be a teaching assistant for a professor of Urban Studies at your university. The professor plans to begin her course with a broad overview of the growth of cities through time. She has provided you with web maps and data showing the top 10 cities (in population) in 100, 1000, 1500, 1800, 1900, 1950, and 2000. She has asked you to convert her maps into a story map that tells the narrative of urban growth and distribution over this span of centuries.

Build skills in these areas

▸ Using the online library of Builder templates to select a template for your story map

▸ Searching ArcGIS Online for maps

▸ Creating a Map Series Web App using the Map Series Builder

What you need

▸ An ArcGIS Online account (free public or organizational)

▸ Estimated time: 1 hour

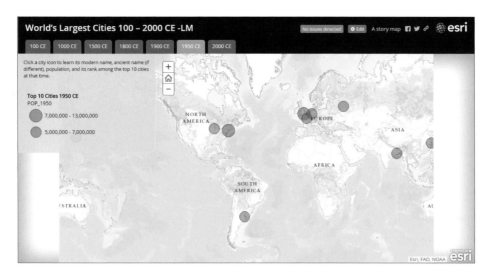

Use a Story Map Builder template to create a Map Series web application

1. Go to the Story Maps Apps page (http://storymaps.arcgis.com/en/app-list/) and select Story Map Series – Tabbed Layout.
2. Click Build.

3. Sign in to your ArcGIS Online account.
4. Click to go to the Map Series Builder.
5. Give your Tabbed Map series a title: **World's Largest Cities 100 – 2000 CE** (put your initials at the end of the title) and click the arrow.
6. Create your first tab.
   a. Title it **100 CE**
   b. In the Map dropdown list choose Select a Map.

   c. In the Select a map window enter "citypop" in the search term box and search ArcGIS Online for a map titled citypop 100. Choose the map with that name and click Add.

d. In the next window, check Legend, and click Add.

e. Enter the following in the empty text box:
   Click a city symbol to learn its modern name, ancient name (if different), population, and its rank among the top 10 cities at that time.

7. Zoom to an extent that includes the Western and Eastern hemispheres.
8. Click Save.
9. Click Add to add another tab.

   a. Name the new tab   **1000 CE**.
   b. Choose a map.

c. In the Select a map window enter "citypop" in the search term box and search ArcGIS Online for a map titled citypop 1000. Choose the map with that name and click Add.

d. Add the same text to the empty text box: **Click a city icon to learn its modern name, ancient name (if different), population, and its rank among the top 10 cities at that time.**

10. Click Save.

11. Click Add to add your third tab.

   a. Name the tab **1500 CE.**

   b. Once again choose to select a map.

   c. Search ArcGIS Online for a map titled citypop 1500, using the search term citypop.

   d. Insert the same text into the empty text box: **Click a city icon to learn its modern name, ancient name (if different), population, and its rank among the top 10 cities at that time.**

12. Continue in the same manner to create the remaining tabs: 1800 CE, 1900 CE, 1950 CE, and 2000 CE.

   a. The maps you will search for are citypop 1800, citypop 1900, citypop 1950, and citypop 2000.

   b. Add the same content to the text box each time.

Your finished map looks like this. Each tab displays the top 10 cities for a particular year.

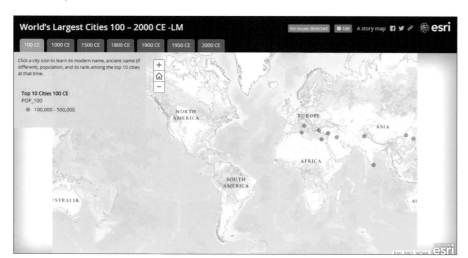

Extended activity
- Create a similar map series focused on U.S. cities between 1700 and 2000.

Resources
- *Four Thousand Years of Urban Growth: An Historical Census* by Tertius Chandler

Lesson 3-4: Create a spyglass story map
Travel through time in Washington DC

Lesson 3-2 introduced us to the value of GIS for exploring potentially related variables, such as unemployment and high school graduation rates or diabetes and obesity rates. We did this by comparing mapped data of the two variables in a swipe map. GIS is similarly invaluable in comparing maps or imagery of the same location from different time periods or in comparing historic maps with contemporary maps and imagery. The power of GIS provides the lenses to make such comparisons and ArcGIS Online makes it easy to share the vision.

Story Maps provide a number of ways to compare maps from different dates. The Story Map Swipe and Spyglass app enables users to interact with two web maps simultaneously in a single scalable view. In this lesson you will create a Spyglass Story Map to compare Washington, DC in 1851 with imagery of the modern capital city. You will add Map Notes to the map to show the three locations for scenes in the play.

Scenario

A new play, set in 1851 Washington, DC, is scheduled to open in that city in six months. The theater owner wants to attract attention to the play and enhance theatergoers' experience by providing a way for them to relate the city they know to the city of the play's setting. They have hired your data visualization company to produce a web map that can be embedded in online promotions for the play and can be accessed through visualization kiosks at various locations around the city and in the theater itself. They have

asked for a map that facilitates easy comparison between the 1851 city and its current configuration, shows key locations in the city that are portrayed in the play's three scenes (National Hotel, Naval Observatory, US Patent and Trademark Office), and enables users to search for specific addresses.

Build skills in these areas

▸ Opening a map

▸ Adding map notes to the map

▸ Creating a Story Map Spyglass web application

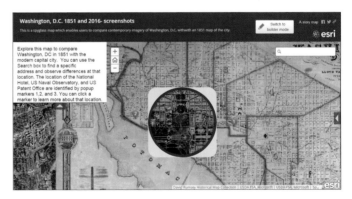

Washington D.C. 1851 and 2016

What you need

▸ An ArcGIS Online account, public or organization

▸ Estimated time: 45 minutes

Open a map displaying an historical map of Washington, DC

1. Sign in to your ArcGIS Online account.
2. Click Search
3. Serach for LearnResource.
4. Click the thumbnail of the Washington D.C. 1851 and 2016 map to open it
5. When you have opened the map use Save As to save it to My Content in your account. Replace the word Copy with your initials (e.g., Washington D.C. 1851 and 2016 LM).
6. Click the Contents tab to see the map's layers. If you turn off the World Imagery layer, you will see the 1851 map beneath it.

Add map notes

1. Follow the steps below to create three Map Notes to show the locations of the National Hotel, Naval Observatory, and US Patent and Trademark Office in 1851. The steps are the same for each note, but the information about each location will differ. Use the table that follows to fill in the information for each note.
   a. Choose Add Map Notes (under Add).

   b. Name the Map Note (National Hotel, Naval Observatory, or Washington Asylum).

   c. Accept the default template (Map Notes) and click Create.

   d. Use the information in these sample tables to fill in the Map Note information options for each site.

Map Note 1	
Location	-77.0193, 38.8931
Title	Scene 1 - National Hotel
Description	The National Hotel was a major landmark for much of Washington's early history. It was built incrementally by John Gadsby between 1827 and 1857.
URL to an image	http://bit.ly/1r6Gp3m
Image Link	http://bit.ly/1qQRipJ
Change symbol	Numeral 1, 25 pts

Map Note 2	
Location	-77.05145, 38.8951
Title	Scene 2 - US Naval Observatory
Description	The U.S. Naval Observatory is the preeminent authority in the areas of time keeping and celestial observing; determining and distributing the timing and astronomical data required for accurate navigation and fundamental astronomy.
URL to an image	http://earlyradiohistory.us/1905tim1.jpg
Image Link	http://www.usno.navy.mil/USNO/about-us/brief-history
Change symbol	Numeral 2, 25 pts

Map Note 3	
Location	-77.0297, 38.8971
Title	Scene 3 - US Patent Office
Description	The Greek Revival U.S. Patent Office building was designed by architect, Robert Mills and constructed between 1836 and 1867. Today it houses the Smithsonian's National Portrait Gallery and the American art Museum.
URL to an image	https://www.census.gov/history/img/PatentOffice.jpg
Image Link	http://1.usa.gov/1VwY3Ko
Change symbol	Numeral 3, 25 pts

e. When you have created each Map Note, click Close and click Edit.

You should now see the three map notes listed in your map Contents.

2. Zoom your map to an extent like the following (you just see the 3 Map Notes).
3. Save the map.

Create a Story Map Swipe web application

1. Click Share (above the map).
2. Click Create a Web App under Embed this map.

3. Click Build a Story Map
4. Select Story Map Swipe and Spyglass, and click Create App.

5. Fill in the New App information box:
   a. Title: **Washington, D.C. 1851 and 2016**
   b. Tags: **Washington DC 1851, spyglass**
   c. Summary: **Spyglass Story Map of an 1851 map of Washington, D.C. and contemporary satellite imagery**
6. Click Done.
7. Click Select this layout under Spyglass, and click Next.

8. Use the dropdown list to choose World Imagery as the layer to appear within the spyglass.

9. Click Next.

10. Choose the following layout settings: enable description, enable popup, and enable an address search tool.

11. Click Next.
12. Enter the following headers on the next Swipe/Spyglass Builder window:
    Main Map Header Title: **Washington, DC 1851**
    Spyglass Header title: **Washington, DC Today**
13. Click Open the app.
    Your app preview looks like the image below.

Configure your Story Map Spyglass web application
1. Enter the following text in the Edit me! Box:
   a. **Explore this map to compare Washington, DC in 1851 with the modern capital city. You can use the Search box to find a specific address and observe differences at that location. The location of the National Hotel, US Naval Observatory, and US Patent Office are identified by popup markers 1,2, and 3. You can click a marker to learn more about that location.**
2. Experiment with different font sizes and families.
3. Edit the app's Title and Subtitle if desired.
4. When you are satisfied with the changes you've made, click Save under Story Configuration. If you go to My Content and open the saved app from there, it looks like this.

5. Explore the app.
   a. Click on one of the map notes symbols to see the popup appear on the side of the map.
   b. Search for a modern street address in the search box.
   c. Move the Spyglass around on the map and zoom in for a closer view.

Extended activity
• Create a Spyglass map to compare two historic maps or imagery from two different dates.

The ArcGIS Book, chapter 3
Questions for reading comprehension, reflection, and discussion

Teachers can use the items in this section as an assignment, an introduction, or an assessment, tailored to the sophistication of the learners. Some learners can read all the sections at one time, while others are more comfortable with small segments. The questions and tasks are designed to stimulate thought and discussion.

1. Story maps: The fusion of maps and stories come to life

   a. List components that can be incorporated into a story map.

   b. Story maps "use the tools of GIS, and often present the results of spatial analysis, but don't require their users to have any special knowledge or skills in GIS." (*The ArcGIS Book,* page 37) Identify two maps in the Smithsonian's *Age of Humans Anthropocene Atlas* story map that illustrate this idea.

2. The world of story maps: A gallery of exceptional examples from around the globe

   a. The chapter opens with a gallery of four story maps. Open each one and write a brief statement (one or two sentences) summarizing the story it communicates.

   • Twister Dashboard:

   • Geography Bee:

   • The Age of Megacities:

   • Exploring China's Burgeoning Highways:

3. Who creates story maps: For the people, by the people

   a. Story maps are created by individuals or groups who want to communicate effectively to "affect change, influence opinion, create awareness, raise the alarm, and get out the news." Explore the six story maps shown here as examples, then complete the chart below to suggest ways that each of these organizations could use story maps to support its own agenda.

Organization	Who is the audience they want to reach? What is the purpose of the story map?	Suggest a story they might want to communicate with a story map.
A local historical or preservation group		
A state health department during flu season		
The campaign staff for a Presidential candidate		
A city or state law enforcement agency		

4. Thought Leader: Allen Carroll: Why maps are so interesting

   a. Maps organize information spatially – what does this mean?

5. Maps tell stories: What kinds of stories can you tell?

   a. Story maps serve many purposes. This section identifies eight kinds of stories that maps can tell – list them and write a brief sentence summarizing what the story map depicts or shares.

   1. Describing places:

   2. Comparing data:

   3. Revealing patterns:

   4. Presenting narratives:

   5. Recounting history:

   6. Celebrating the world:

   7. Breaking news:

   8. Depicting change:

   9. Based on your own interests and knowledge, briefly describe two or more stories you'd like to tell with a story map

6. Quickstart: Combine your maps and customize interactive apps to tell a story.

Planning is very important when creating a story map. Pick one of the stories you just indicated you would like to tell with a story map and develop a plan for creating it by briefly writing the answers to the following questions:

   a. What is your purpose and goal in telling the story and who is your audience?

   b. Go to the Story Maps Gallery and identify two or more story maps whose subject matter is similar to the story you'd like to tell.

   c. Browse story map templates on the Story Maps Apps and identify the one that seems best for your story map project.

Additional resources

ArcWatch, January 2016, Create Story Feature Helps You to Choose the Right Story Map, http://www.esri.com/esri-news/arcwatch/0116/create-story-feature-helps-you-to-choose-the-right-story-map

Maps We Love, http://www.esri.com/products/maps-we-love

Twelve Days of Story Maps Tips, ArcGIS Resources blog, https://blogs.esri.com/esri/arcgis/2015/12/21/twelve-days-of-story-map-tips/

Nine Things You didn't Know You Could do With Story Maps, ArcGIS Resources blog https://blogs.esri.com/esri/arcgis/2016/02/16/story-maps-9-things/

More online lessons

At the end of chapter 3 in the online version of *The ArcGIS Book* (http://learn.arcgis.com/en/arcgis-book/chapter3/), you'll find lessons to guide you through the creation of two more story maps using the Story Map Tour and the Geoportfolio apps. Each of these lessons can be completed with an ArcGIS Public Account. Both provide practice in integrating photographs and smartphones into the story map-making process.

Great Maps Need Great Data
Tap ArcGIS Online and the Living Atlas of the World for crucial information

You are meeting friends at a restaurant you've never been to, so you do an Internet search for it, click on its map location and then the "Get directions" link to see how to get there, and in less than a minute, you're good to go! We don't think twice about this because, it's such an everyday thing. Yet it's an indication of a remarkable era of transformation, and we're right in the midst of it. Not so long ago you would have had to consult a phone directory and a local map to determine where the restaurant was and how to get there, and call your friends to relay the directions. You'd speak to them, they'd take notes—that was what using and sharing data meant in the relatively recent past. But now, though there's less to do, there's so much more to it—and so much more to do with it.

The transformative medium of "cloud" data storage means we now have immediate access to data, and not just about local restaurants and streets but data about demographics, land cover, urban infrastructure, and much more. This is the kind of information that, when you think about it (which GIS also helps you do), becomes knowledge with which to solve problems and make decisions about vital issues. With such widely available "open" data, anyone can ask and answer geographic questions quickly. We can create and share maps integrating authoritative with local or regional data. ArcGIS Online is rapidly emerging as the technology of choice for using spatial data to explore and work out solutions for real-world problems. It truly is a whole new world.

This chapter provides guided practice in strategies for using open data to answer geographical questions and solve problems. Two scenario-based lessons demonstrate the power of open data to support informed decision making. Additionally, the step-by-step lessons build and reinforce fundamental GIS skills. Use the questions at the end of this chapter to support your reading comprehension, reflection, and discussion of the narratives presented in the corresponding chapter 4 of *The ArcGIS Book*.

Introductory activities

Video
Go to geospatialrevolution.psu.edu/episode2
The Geospatial Revolution, Episode 2, produced by WPSU Penn State

This episode begins with a look at how the City of Portland, Oregon, is using geospatial technology to provide its citizens with information they can use to make decisions and live their lives. How is open data making Portland a more livable city?

Activity
Welcome to MapQuiz by Esri
Go to http://maps.esri.com/rc/quiz/index.html

One component of ArcGIS Open Data, now accessible via the web, is imagery: photographic, satellite, and multispectral. This engaging quiz utilizes photographic imagery to test your geographic literacy. Sign in with a Facebook account to get started.

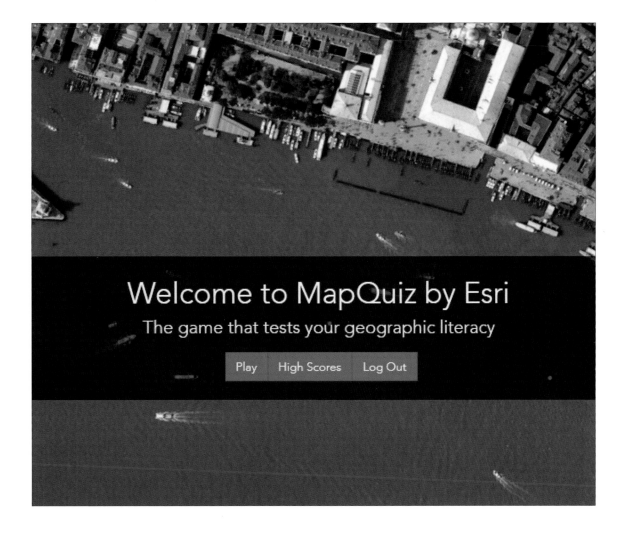

Welcome to MapQuiz by Esri
The game that tests your geographic literacy

Play   High Scores   Log Out

Lesson 4-1: Map US Minority Populations

Acccessing open data

Map US minority populations

One of the oldest sources of open data is the US Census Bureau. Decennial census data was always available, but it was not until the 1990 Census that data for the entire United States became available in a digital format designed for use in a GIS. That year saw the release of the first nationwide digital map of the United States, called the TIGER (Topologically Integrated Geographic Encoding and Referencing) database. Today census data is used for research, program design, and planning in every segment of society (government, business, health, environment, education, and many more).

Census data can be mapped and analyzed at every geographic level from states to census tracts. In this lesson, you will map the distribution of minority populations across the United States by county. After creating your map, you will share it as a Bulleted Series Story Map.

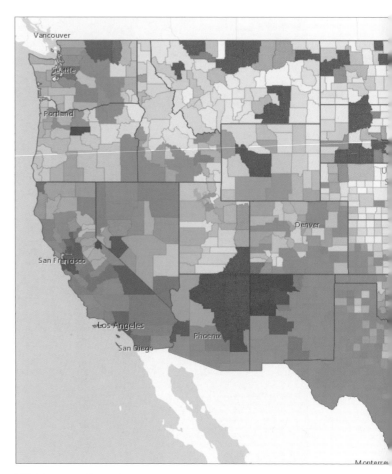

US Minority Populations

Scenario

The US Department of Education is undertaking a study of minority student performance on standardized achievement tests. They have hired your Data Visualization Company to produce maps showing the distribution of minorities in the United States. Your company has chosen to use ArcGIS Online to produce the required product. The Census Bureau has given you data for minority population by county. The DOE has asked for a map symbolized to show minority population as a percent of total population and to specifically identify counties in which the minority population is 50 percent or higher.

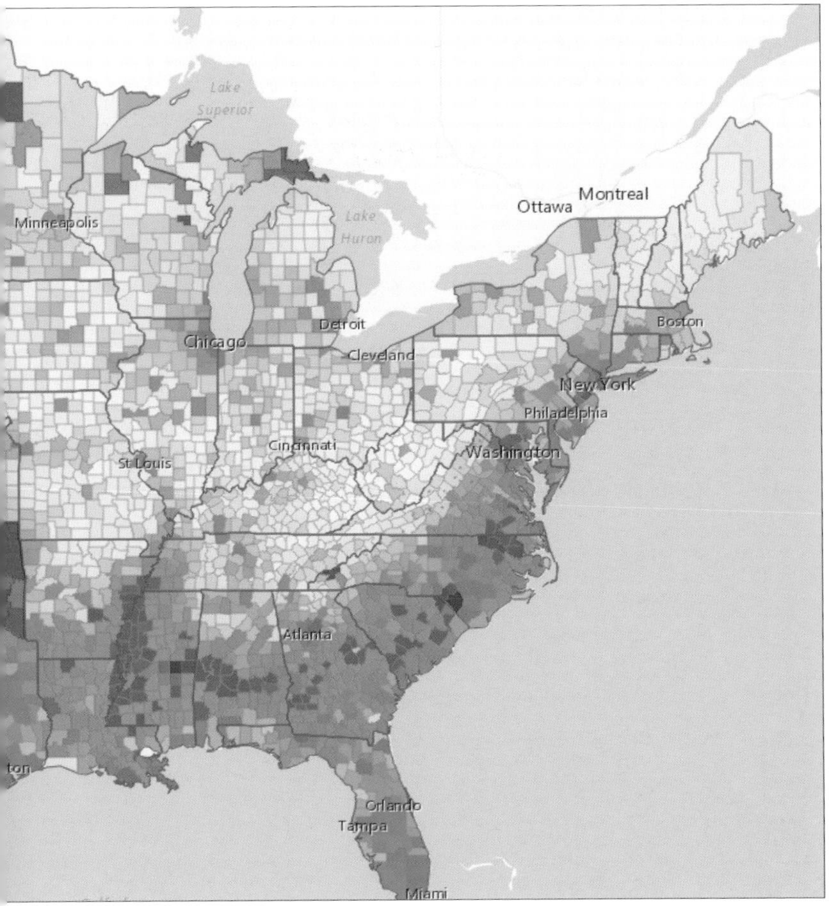

Build skills in these areas

▸ Opening a map

▸ Making copies of a map layer

▸ Renaming map layers

▸ Classifying minority populations by percent of total county population

▸ Creating layers showing counties with minority populations greater than 50%

▸ Creating a Bulleted Series Story Map

What you need

▸ Account not required

▸ Estimated time: 1 hour 30 minutes

Open the map
1. Go to http://arcgis.com.
2. Click Sign In.
3. Click Search.
4. Search LearnResource.
5. Click Minority Populations by County.

Make copies of a map layer

1. In the list of map contents, click More Options (3 dots) under USA Counties and select Copy.

You now see a new layer called USA Counties – Copy.

2. Repeat the copy step until you have 5 layers called USA Counties – Copy.
3. Click More Options (3 dots) under USA States and click Move up.
4. Repeat until the USA States is the top layer in the Table of Contents.

You can now see the state boundaries as well as the county boundaries and your map should look like the one below.

5. Save your map as **Minority Populations by County [your initials]**.

Rename map layers
1. Click More Options (3 dots) under one of the USA Counties – Copy layers and select Rename.
2. Enter **Black** in the layer name box that appears.

3. Repeat this process naming the other USA Counties – Copy layers Hispanic, Asian, Native American/Alaskan, and Hawaiian/Pacific Islander (the sequence doesn't matter).

Your Table of Contents should now look like this:

Classify minority populations by percent of total county population

1. Click Change Style under the layer named Black.

Leave the attribute selection as Show location only and click Options to change the symbol to a bold color.

2. Click Attribute Values under the Transparency slider. This means you want to set the transparency by an attribute value.
3. Select black from the dropdown list of attributes and Pop2010 from the Divided By list.
4. You see a window that says Set transparency based on the attribute values. Make the following changes in this window:
   a. Under Transparency Range, change High Value to 0% and Low Value to 100%.
   b. Change the breakpoint on the slider from .22 to .2 (you can do this by clicking on the .22 and typing in the new number).
   c. Click OK, click OK, click OK, and click Done.

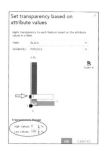

5. Turn off all the minority group layers except Black and USA States. Your map now looks like this:

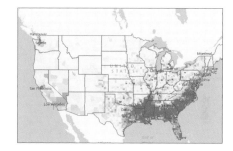

6. Repeat the same procedure for Hispanic, Asian, Native American/Alaskan, and Hawaiian/Pacific Islander.
7. Save your map frequently.

When you have symbolized all the layers, your map will look as it does here. You need to pan to see Alaska and Hawaii. The appearance of the map may vary, of course, depending on your choice of colors and the sequence of layers from top to bottom.

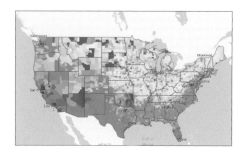

Create layers showing counties with minority populations greater than 50%

1. Copy the USA Counties layer.
2. Rename it Black 50% or higher.
3. Click Change Style under the new layer and select Black from the dropdown menu under Choose an attribute to show.
4. Click Options in the Counts and Amounts (Color) drawing style.
5. In the Divided By dropdown list, select POP2010.

6. Under classify Data, select 2 classes.
7. On the slider, change the breakpoint to .5.
8. Click Legend and change the 0 to 0.5 to No fill.
9. Click OK and click Done.

The Black 50% or higher layer should look like this.

10. Repeat this procedure two more times, creating a Hispanic 50% or higher layer and a Native American/Alaskan 50% or higher layer. (If you try to follow the same procedure with the Asian or Hawaiian/Pacific Islander layers you will get a message "This value is out of range" when you try to change the breakpoint to .5. This means there are no counties in which these minorities make up 50% or more.)

The three layers showing counties with minority population at or above 50% should look like this:

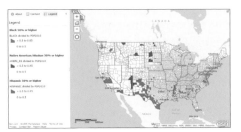

You will need to save a separate map of the counties with 50% or higher minority population.

11. Turn off all the minority layers except for the ones showing the counties with 50% or higher minority population.
12. Use Save As to save this map as Counties: Minority Majority.

Create a Bulleted Series Story Map

1. Open your first saved map: Minority Populations by County [your initials].
2. Select Share and Create a web app.
3. Select Build a Story Map and click Story Map Series.
4. Title the app **Minority Populations by County – [your initials]**. Give it tags and an app summary and click Done.
5. In the Map Series Builder window, select Bulleted and click Start.

6. Enter the title of your Bulleted Map Series and **Minority Populations by County** and click the arrow.
7. Give your Side Bulleted Map series the title **Minority Populations by County**, and click the arrow.
8. Create the first entry: **Entry title – U.S. Minority Populations.**
9. Choose Select a Map, browse the maps in your account to find Minority Population by County, and click Add.

10. This opens a Builder window for your first entry. Enter the following text in the Text area:

This map reflects the distribution of minority populations, by county, across the United States and is based on 2010 Decennial Census data. The darkest color for any group indicates that it represents 20% or more of the total county population.

11. When you are finished, click Save (upper right).
12. Click Add (upper left) to add another entry.
13. Title the new entry **Minority Population 50% or Greater.**
14. Choose Select a Map, browse the maps in your account to find Counties: Minority Majority, and click Add.
15. This opens a Builder window for your second entry. Enter the following text in the Text area: **This map shows U.S. counties in which the minority population is 50% or more of the total county population**
16. Save your story map once again.

Explore options in the Settings area. If you wish to do so, you can change the size of the legend and text panel, add tools such as a location finder, and set the theme colors of the app. Once you have made changes through the settings option, save your map a final time. As the app owner, you can always change these items again later.

Lesson 4-2: Help restore a watershed
Chesapeake Bay States Land Use Enrichment

The Chesapeake Bay is the largest estuary in the contiguous United States. The watershed covers 64,000 square miles and is fed by 50 major rivers and streams. Because everything always flows downstream, all the results of human habitation run into the ocean. The category of human habitation includes farms and developed lands, particularly when you consider that as such land is farmed and developed, there is less soil to absorb and filter the water. Restoration efforts have made modest ecological gains but they have been largely offset by rapid population growth. Agriculture and developed land is one of the land use classifications that are used to study the health of the Bay.

Scenario

The Chesapeake Bay Foundation is interested in correctly assigning funds and resources to the seven states that make up the Chesapeake Bay watershed. In order to do this they have asked for both quantitative and qualitative information from the GIS department. They want to know what percentage of land use is in each designated state's watershed area and they want it displayed in a compelling way to present to the community.

Build skills in these areas

▸ Opening a map

▸ Enriching a layer

▸ Configuring Popups

▸ Using Popup Media

▸ Calculating Percentage

▸ Publishing as a web app

What you need

▸ User, Publisher, or Administrator role in an ArcGIS organization

▸ Estimated time: 1 hour 30 minutes

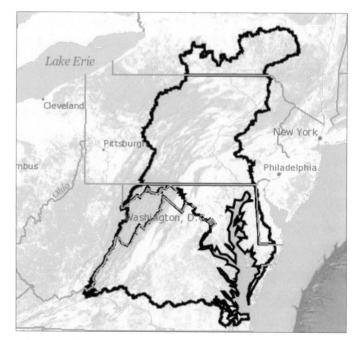

Chesapeake Bay Watershed Land Use Enrichment

Open the map

1. Sign in to your ArcGIS organization account.
2. Search for the Chesapeake Bay Landuse ArcGIS Online group.

3. Uncheck Only search in Participants and Resources.
4. Click the Chesapeake Bay Landuse group.

5. Click the thumbnail of the Chesapeake Bay Landuse map and click Content under Add.

The map opens showing the Topographic basemap, the outline of the Chesapeake Bay, and the section of each state in the Bay.

Save the map

1. On the top of the page click Save and choose Save As.
2. In the Save Map window, type **Answers: Chesapeake Bay Landuse**.
3. Type tags and a brief description of the map's content, and Save map.

Display the Bay Area by State abbreviation

The Bay Area by State layer is s displayed without any cartographic styling.

1. Click show Contents of Map below Details.
2. Click Change Style below Bay Area State layer.

3. The states are shown by location only. Choose STATE_ABBR in the Choose an attribute to show tab.
4. Click Done. The states are now displayed by

individual colors.

a. Write several spatial observations about the Chesapeake Bay watershed. Include in your observations a discussion of political vs natural boundaries. The observations that you write here will be used later in the exercise when you construct your story map. You might want to save your answer in a word document so you can use it later in the exercise.

Use the Data Enrichment tool

The Data Enrichment tool helps you explore the character of areas. Detailed information is returned for the chosen area. For this area you would like to know information about the percentage of land use or land cover in each of the states represented in the Chesapeake Bay.

1. Click Perform Analysis.

2. Click Data Enrichment.
3. Click the Enrich Layer tool to open the Enrich Layer menu.

4. The Bay Area by State is selected by the layer to enrich. Click Select Variables to choose the land cover information you want to retrieve.
5. In the Data Browser, in the upper left, be sure that United States is selected.
6. Click the bottom button to show page 2.
7. Click Landscape.
8. Click Landcover.

9. Select all seven variables by checking the Landscape Analyst Variables.

10. Click Apply.
11. The Result layer name should be Enriched Bay Area by State.

12. Uncheck Use current map extent.
13. Click Run Analysis. The Enriched Bay Area by State layer is added to the Contents pane.
14. Click Show Table.

When you open the table, you see that each bay state has been populated with the percentage of land use.

15. When you examine the table, you see several fields that are not necessary for the analysis. Click Table Options on the right and click Show/Hide Columns.
   Uncheck the fields listed below:
   a. STATE_NAME
   b. ENRICH_FID
   c. ID
   d. sourceCountry
   e. HasData
   f. aggregatedMethod

16. Close the table by clicking the X in the right corner.
17. Uncheck Bay Area by State in the Contents pane.
18. Use the Change Style icon to display Enriched Bay Area by STATE_ABBR.
19. Click Done.

Configure attributes and show charts

You can display the bay states by land use percentage, but the amount of information that you portray is neither compelling nor quantitative. A better way to present your data is to create a chart. A chart will graphically display the values of numeric attribute fields, in this case the percentage of land use.

Before you chart your data you need to configure popups.

1. Click the three buttons on Enriched Bay Area by State and go to Configure Pop-up.
2. Enter a title for your popup: **Title = Enriched Bay Area by State**.

3. Select an attribute option from the dropdown menu. Select A list of field attributes.
4. Click Configure Attributes and uncheck the following:
   a. STATE_NAME
   b. ENRICH_FID
   c. ID
   d. sourceCountry
   e. HasData
   f. aggregatedMethod
5. Click within the attributes and change to the following:

a. STATE_ABBR
b. % Open Water
c. % Barren Land
d. % Developed
e. % Forest
f. %Herbaceous
g. %Pasture/Crops
h. %Wetlands
6. Click OK.
7. Close the table by clicking the X in the upper right corner.

8. Under Pop-up Media click Add and add a Pie Chart.
9. Configure Pie Chart as follows:
   a. Title = Chesapeake Bay Land Use
   b. Caption = % of Land Use by State
   c. Check the following:
   - % Open Water
   - % Barren Land
   - % Developed
   - % Forest
   - %Herbaceous
   - %Pasture/Crops
   - %Wetlands
10. Click OK.

Now when you click on each of the states it shows the percentages of land use in both a table and a pie chart.

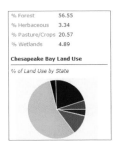

Use the table

Don't forget the interactive table is a quick way to analyze information and make decisions. Sorting by a specific attribute is always useful.

1. Show the table.
2. Click the field % Developed.

3. Click the gear and choose Sort Descending. That puts the values in order from highest to lowest and you can see that DC is by far the most developed state followed by MD and VA.

STATE_ABBR	% Open Water	% Barren Land	% Developed		
				Sort Ascending	
				Sort Descending	
DC	3.47	0.08	80.76	Σ Statistics	
MD	6.73	0.31	17.65	Calculate	
VA	3.55	0.18	10.93	Delete 56.55	3.34

You can also sort the table to arrive at more quantitative data.
4. Close the table.
5. Click Save.

Create web app

The Chesapeake Bay Foundation asked you to show the percentage of land use by each state in the bay area in a compelling way to present to the community. You have decided to create a story map to fulfill this task. You have selected a story map that presents a series of maps and other content organized by using tabs.

You have decided that the individual state maps you want to use in your story map will look better if the basemap is Imagery with Labels.

1. Change the basemap to Imagery with Labels and save it.

2. Click Share and Share with Your Organization. You need individual maps of each watershed for your story map.
3. Filter the map by STATE_ABBR and select NY. Save the map as NY and share with Everyone.
4. Repeat steps 1 through 4 for all the states. You should have seven maps.
5. Close the individual state maps.

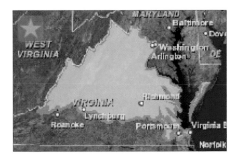

6. Open: Chesapeake Landuse.
7. Click Share.
8. Click Create App.
9. Click Build a Story Map.
10. Click Story Map Series.
11. Click Create App.
12. Specify a title, tags, and a summary for the new web app.
13. Click Done.
14. Select Tabbed on the Welcome to Map Series Builder.
15. Click Start.
16. Type .**Chesapeake Bay Watershed** as the title for your Tabbed Map series.
17. Click the arrow.
18. Add Land Use Analysis for the Add Tab.
19. Click a map pull-down menu and choose Chesapeake Bay Landuse.
20. Click Add.

21. At the top of the page, click Settings.
22. Click Map options.
23. Uncheck Synchronize map locations.
24. Click Apply.
25. Add some appropriate text about the Chesapeake Bay to the text box.
26. Click Add.
27. Select a map. (Select NY first.)
28. Click Add.
29. Click on the bay part of New York, copy the percentages, and add to your text box.
30. Repeat steps 28 through 30 for the other six states.
31. Click Save in the upper right corner.
32. Click View Live to review your story map.

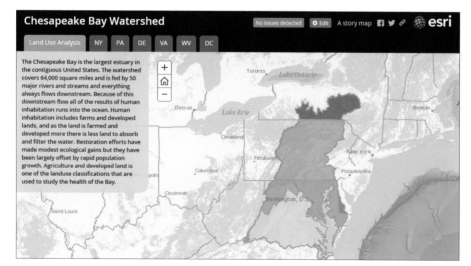

More online lessons
At the end of chapter 4 in the online version of *The ArcGIS Book* (learn.arcgis.com/en/arcgis-book/chapter4/), you'll find the Learn ArcGIS Lessons to create a map using landscape layers from the ArcGIS Living Atlas as well as an elevation profile map. To do each of these lessons you must be logged in to an organizational account.

The ArcGIS Book, chapter 4
Questions for reading comprehension, reflection, and discussion

Teachers can use the items in this section as an assignment, an introduction, or an assessment, tailored these to the sophistication of the learners. Some learners can read all the sections at one time, while others are more comfortable with small segments. The questions and tasks are designed to stimulate thought and discussion.

1. The Living Atlas: The ArcGIS platform provides rich content

   a. How have open data repositories like the Living Atlas changed the way GIS users plan and implement projects?

2. The ArcGIS data community: A global network for creating and sharing authoritative geographic information resources

   a. The work of organizations that make up the global GIS community has changed with the emergence of web GIS. Explain how this GIS work has changed.

3. What kind of data is available? Definitive, authoritative basemaps

   a. There are several key concepts to understand about basemaps; they are multiscale, provide global coverage, and are continuous. In your own words, briefly define each of these concepts.

Multiscale:
Global coverage:
Continuous:

4. Demographics

   a. Explain the concept of data enrichment.

5. Opening data to the world of possibilities

   a. What are Open Data sites and what benefits do they provide?

6. Imagery

   a. In your own words, briefly compare these types of imagery: photographic, satellite, and multispectral.
   Photographic:
   Satellite:
   Multispectral:

7. Landscapes: Landscape analysis layers

   a. What are ELUs?

8. Thought Leader: Richard Saul Wurman: A map is a pattern made understandable

   a. In your own words, explain what the phrase "understanding precedes action" means to you.
   b. Explore the Urban Observatory and explain how it reflects the concept that "understanding precedes action."

The Importance of Where
Answer complex questions using ready-to-use maps and analysis tools

Once data is mapped, a unique set of tools available in ArcGIS Online can be used to perform analysis. These spatial tools are extensive and they take you to the heart of why GIS helps you delve deeper into any project or inquiry before you. There are tools to summarize and aggregate data based on a geographic entity, network and locational tools, surface analysis tools, visibility analysis tools, tools that combine layers of data, and statistical tools, to list a few. They may sound out of reach to you now, but they are not. Taken in hand and applied to spatial data, they actually extend your reach and enable you to answer questions and solve problems, faster and more accurately than almost anyone could do without them.

This chapter offers the opportunity to do the following:
- Use the spatial problem-solving method of ask, calculate, interpret, decide, and communicate.
- Map and analyze a real-life example of GIS as used for emergency management.
- Analyze crime data using appropriate analytical tools.

Use the questions at the end of this chapter to support your reading comprehension, reflection, and discussion of the narratives presented in the corresponding chapter 5 of *The ArcGIS Book*.

Introductory activities

Video
Go to video.esri.com and search for the video by title.

UC2015 Opening Video
Applying Geography
A compelling video about the planet, humans, and applying geographic knowledge.

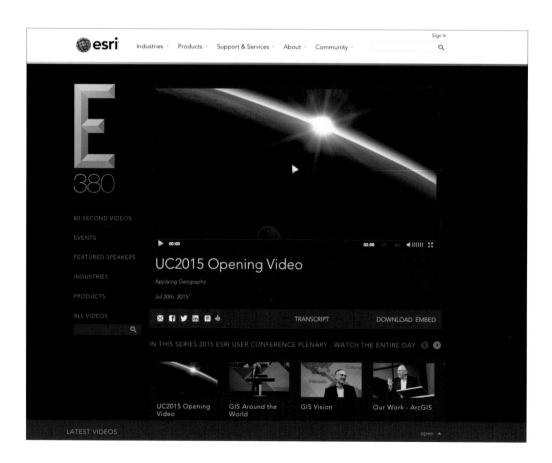

Activity
Ask, calculate, interpret, decide, communicate
Find the *Water Quality* and *Conservation* maps, which you will use for this lesson, in chapter 5 of thearcgisbook.com, located under the How is spatial analysis used? heading. *Water Quality* is the second map and *Conservation* is the fifth map.

The spatial analysis process always starts with a question. The question drives the analytical investigation. After asking the question, there is a standard protocol for solving spatial problems. The protocol workflow is: ask, calculate, interpret, decide, and communicate. Access each of the maps below and record what is happening throughout the protocol workflow.

For each map chose below list each of the steps: ask, calculate, interpret, decide, and communicate.

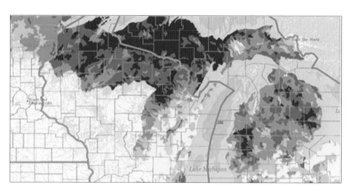
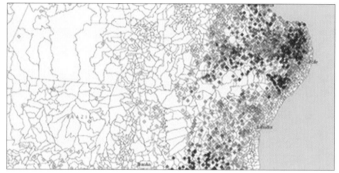

Water Quality

1. Ask:

2. Calculate:

3. Interpret:

4. Decide:

5. Communicate:

Conservation

1. Ask:

2. Calculate:

3. Interpret:

4. Decide:

5. Communicate:

Lesson 5-1: Analyze a severe windstorm

Spatial analysis and emergency management

This section provides a real-life example of GIS being used for emergency management. You are tasked with creating maps that would be used by first responders or the general public during an emergency situation. You will use symbolizing/classification and querying/filtering along with the geospatial tools of proximity and overlay.

To work on this lesson, you do not have to be logged in to an organizational account, however if you're not you cannot save your work. Here are some suggestions as to how to deal with this:
- Block off a long period of time.
- Assess the work in parts.
- Take screenshots of your work to show completion to your instructor.

Emergency management unfolds at all levels and benefits from a technology designed to make spatial decisions. Maps can be produced for first responders and decisions can be made for resource management. Because of these benefits there is a growing demand for the use of GIS at all levels of emergency management. Thought leader Linda Beale summarizes it best: "GIS offers the technology to explore, manipulate, analyze, and model data from multiple sources. With spatial analysis hazard mapping and predictions developed for risk assessment, you can use models to evaluate response strategies, and maps to illustrate preventative strategies and for risk communication and negotiation."

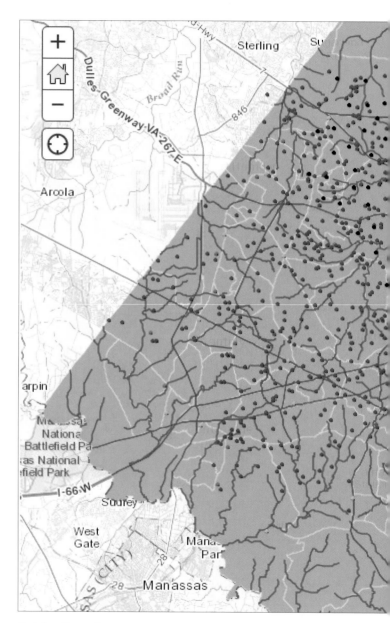

Fairfax Derecho

Scenario

A derecho is a widespread, long-lived, straight-line wind storm that is associated with a land-based, fast-moving group of severe thunderstorms. At 10 p.m. on June 29, 2012, a derecho hit Fairfax County, Virginia. Within one hour the County's Emergency Operations Center was activated and additional agencies from the public schools, facilities management, park authority, Virginia Department of Transportation, Dominion Power, and local phone companies were contacted. The Fairfax County GIS department was called to provide maps of the situation. These maps were to be made available online to first responders and the general public. As a GIS analyst you are in charge of a map that shows the following:

- Census Tracts displayed by population density.
- Hospitals, schools, and fire stations symbolized.
- Symbolize rivers and highways.

Build skills in these areas

▸ Symbolizing

▸ Classifying

▸ Creating map notes layers

▸ Using advanced filtering

▸ Generating a heat map

▸ Generating a density map

▸ Overlay: Identifying high-risk intersections

What you need

▸ Account not required

▸ Estimated time: 2 hours

Part 1: Opening the map and classification/styling

1. Go to http://arcgis.com
2. Click Sign In.
3. Click Search.
4. Search for LearnResource.
5. Click the Fairfax Derecho icon to open the map.

6. In the upper right corner click Modify Map.
7. Click Show Contents of Map under Details on the top of the page. This reveals all the layers that are available for your analysis.

8. Check Census Tracts.

The census tracts are shown all as one color or by location only. The request from the Emergency Center was to display the census tracts by population density in 2010. This will show where there is the most concentration of people. The attribute to choose is POP10_SQMI which indicates the number of people per square mile in 2000.

9. Click Change Style under Census Tracts.

10. Choose POP00_SQMI for the attribute and click Counts and Amounts (Color).
11. Click Done.
12. Click Census Tracts in the Contents pane to expand the legend. Notice the density varies from 10,011 people per square mile to 0 people per square mile.

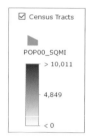

You can definitely see that there are sections of the county that have more people per square mile than others.

13. Click Census Tracts again to close the legend.

Part 2: Symbolization

The map will be easier to read and interpret if it is symbolized correctly. Rivers, highways, schools, hospitals, and fire stations all need to be symbolized.

1. Uncheck the classified or styled Census Tracts layer.
2. Check Rivers to turn on the layer.

The rivers first paint into the map at an appropriate size. It would be better if the rivers were blue to indicate water.

3. Click Rivers and click Change Style.
4. Click Options.
5. Click Symbols and choose an appropriate color.
6. Click OK.
7. Click Done.
8. Repeat steps 3 through 7 for Highways. Choose an appropriate color.

You need to appropriately symbolize schools, hospitals, and fire stations.

9. Click Schools and click Change Style.
10. Click Options.
11. Click Symbols.
12. Click People Places and choose a symbol to represent a school.
13. Change the Symbol Size to 12.
14. Click OK.
15. Click Done.

16. Repeat steps 9 through 15 for Hospitals. Look under the category Safety Health for more symbols.
17. Repeat steps 9 through 15 for Fire Stations. Look under the category Safety Health for more symbols.

Part 3: Queries to find highway intersections where live wires are down

It has been identified that live wires are down at the following intersections:
- Georgetown/Leesburg
- Leesburg/Dranesville
- Dolley Madison/Capital Beltway

You need to first find the intersection and mark with a map note.

1. Turn off by unchecking all layers except Fairfax County and Highways.

Filter highways to identify the locations of the intersection. The first intersection to be identified is Georgetown and Leesburg.

2. Click Filter on the Highways Layer.

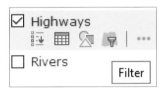

3. You need two expressions from those shown below:
- Name is Georgetown. You must click Unique.
- Add another expression.
- Name is Leesburg.
- Click **Any.**
- Click Apply Filter.

Mark the location of the intersection with a map note.

4. On the top of the page, choose Add and Add Map Notes.

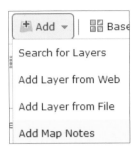

5. Name the Map Note Georgetown and Leesburg.
6. Click Create.
7. Choose the Pushpin and place it on the intersection.
8. Title = Intersection of Georgetown and Leesburg.
9. Description = Live wires are down.

10. Click Close.
11. Click Details.
12. Click Highways and Filter.
13. Click Remove Filter.
    The hidden roads appear.
14. Repeat steps 2 through 14 for the Leesburg/ Dranesville intersection. Remember Any.
15. Repeat steps 2 through 14 for the Dolley Madison/Capital Beltway. Remember Any.

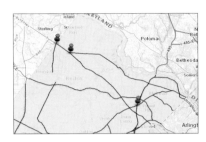

Part 4: Identification of Deployed Fire Stations with a drive time around 10 minutes each
1. Check Fire Stations.
2. Filter using the following expressions:
   - Station is Fox Mill.
   - Station is Springfield.
   - Station is Wolftrap.
   - Check Any.

If you are not logged into an organizational account you cannot do advanced analysis or drive-times. The drive-time area has been made and all you do is check it. A drive-time area is the area that can be reached with a specified drive time. It is a proximity analysis.

3. Check Travel from Fire Stations (10 Minutes).
4. You might want to click Other Options and Move Fire Stations above the Travel from Fire Stations (10 minutes).

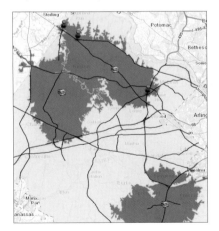

Part 5: Locations of where roads cross rivers

1. Turn off all layers but Fairfax County, Highways, and Rivers.

Intersect is one of the overlay geospatial analysis functions. Because all the layers are bound together geographically these analytical functions work.

2. Turn on the Intersect Highways and Rivers layer. This layer shows where the highways cross the roads.

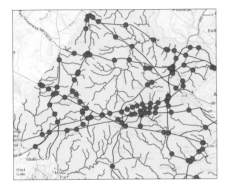

Part 6: Continuous surface maps from point data

The Command Center would like a continuous surface map made from both the phone outages and the power outages. Heat maps show an occurrence of a set of points as a continuous density surface. The density calculation provides a graphical visualization (low to high) and is calculated using a Gaussian blur mathematical function. A heat map has all the following characteristics:

- It is scalable.
- It has no units.
- No data is formed.
- It is not to be confused with temperature.

1. Uncheck all layers except Highways and No Power. To do more analysis you might want to turn other layers, like hospitals and fire stations, on and off.
2. Click No Power and click Change Style.
3. Select Heat Map.
4. Click Done.

You might want to change Basemap to Imagery and turn on Hospitals, Schools and Fire Stations. Notice that as you zoom in and out the heat map is scalable, allowing you to investigate the data at all scales.

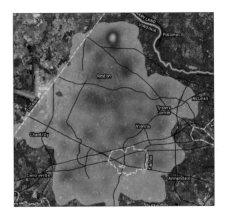

5. Repeat steps 3 and 4 for Phone Outages.

Lesson 5-2: Fight crime in the city
Find patterns and resources in Lincoln, Nebraska

With the introduction of online tools, GIS crime mapping and analysis is being used by law enforcement agencies to make critical decisions and provide their officers in the field valuable information. Law enforcement can now map crime density, look for patterns, and more wisely distribute their resources and personnel.

Scenario

The Lincoln Police Department has committed resources to training, personnel, and software to use GIS to analyze crime in its city. The department wants to look at crime trends and patterns by analyzing different geographical units, such as districts, proximity to highways, and ZIP Codes. They are particularly interested in this analysis because they want to reallocate resources after analyzing the crime patterns. One of their police stations, Team Station, has been targeted as particularly vulnerable to lack of resources. There is also some interest in doing a temporal analysis of crimes by days of the week. You have just been hired by the department to perform this analysis.

Build skills in these areas

▸ Opening a map

▸ Symbolizing data

▸ Using proximity tools: buffer and drive times

▸ Filtering data

▸ Summarizing data of various geographical areas

▸ Creating heat maps

▸ Creating density maps

What you need

▸ User, Publisher, or Administrator role in an ArcGIS organization

▸ Estimated time: 1 hour 30 minutes

You can complete this lesson at learn.arcgis.com. Open the lesson titled *Analyze Patterns of Crime in Lincoln, Nebraska.*

More online lessons

At the end of chapter 5 in the online version of *The ArcGIS Book* (learn.arcgis.com/en/arcgis-book/chapter5/), you'll find a Learn ArcGIS Lesson looking for a suitably zoned area for young adults.

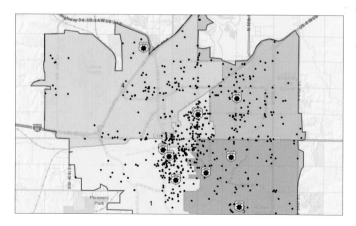

The ArcGIS Book, chapter 5
Questions for reading comprehension, reflection, and discussion

Teachers can use the items in this section as an assignment, an introduction, or an assessment, tailored to the sophistication of learners. Some learners can read all the sections at one time, while others are more comfortable with small segments. The questions and tasks are designed to stimulate thought and discussion.

Spatial problem solving

1. Write a paragraph explaining this statement: GIS is more than a map.

2. Explain one tool from each of the following spatial analysis tools:

   Understand places

   a. Attribute queries:

   b. Spatial queries:

   c. Proximity analysis:

   Detect patterns

   d. Density analysis:

   e. Cluster analysis:

   Determine relationships

   f. Attribute joins:

   g. Spatial joins:

   h. Overlay analysis:

Make predictions

i. Interpolation:

j. Regression:

k. Surface analysis:

Find Locations

l. Site suitability:

m. Location-allocation:

n. Cost corridors:

3. Ask, calculate, interpret, decide, and communicate

    a. How is spatial analysis used? Ask questions and derive answers.

4. Spatial data and spatial analysis

    a. What is the difference between discrete and continuous data?

5. Visualization: What can my map show me?

    a. Why is scale important?

    b. How do style and attributes affect visualization?

    c. Why does the classification scheme you use matter?

6. Explore: What can my data tell me?

    a. Define descriptive statistics:

    b. What are queries?

    c. What is proximity analysis?

    d. Define spatial patterns:

7. Thought Leader: Linda Beale: The challenge is making complex data understandable

    a. Explain the specific role GIS plays in health analysis.

8. Modeling: What can patterns tell you about the following?

a.   Modeling processes:

b.   Interpolating values:

c.   Modeling spatial interaction:

Additional resources

Location, National Geographic Education
http://education.nationalgeographic.org/
encyclopedia/location/

Mapping the Third Dimension
Solve problems in the world as you see it

To be a perfect graphic representation of the world, a map would have to be a sphere. Viewing it in 3D makes it so. The expectation for data to be viewable in 3D is generational. Introducing Esri's ArcGIS Scene to a mixed-age group, you find the older generation becoming a bit awed that they can manipulate, turn, and tilt data in a 3D setting. Anyone under ten is not impressed. Young learners live in a world of 3D and expect data to be displayed in 3D; after all, they have been playing sophisticated 3D games for years.

What does 3D data bring to the table? It introduces vertical and volumetric information. 3D data is a tool for city planners and urban designers, and volumetric analysis serves those investigating things like groundwater contamination. In Esri's ArcGIS Scene, you are able to not only see your data in 3D but also navigate around and through the data at both a local and a global level.

This chapter gives you a look at mapping the third dimension through the following activities:

- Reflecting current events combined with 3D geospatial technology
- Making a 3D presentation
- Visualizing time zones.

Use the questions at the end of this chapter to support your reading comprehension, reflection, and discussion of the narratives presented in the corresponding chapter 6 of *The ArcGIS Book*.

Introductory activities

Videos

Go to chapter 6 of thearcgisbook.com and find the video link under the heading Thought Leader Nathan Shephard, the rise of the 3D cartographic scene. Click Video: *How to author web scenes using ArcGIS Online.*

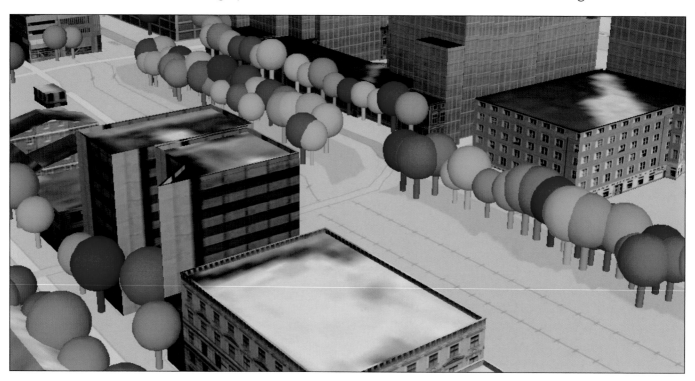

Activity

As with any new app, there are new descriptive terms. Investigate the following highlighted 3D maps and write a brief explanation about the given term and how it is portrayed in the scene. For each part of this activity, you will find the corresponding map in chapter 6 of *The ArcGIS Book*.

1. Find the *Geography and the Assassination of President Kennedy* and Airflow Globe maps in chapter 6 of *The ArcGIS Book* (thearcgisbook. com). Both maps are located under the heading Important 3D terminology. Geography and the Assassination of President Kennedy is a movie location under the subheading Local and global. Airflow Globe is located under the subheading Maps and scenes.

Geography and the Assassination of President Kennedy

Airflow Globe

 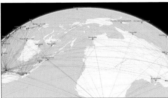

    a. Explain the two different scene environments.

2. Find the Montreal, Canada Scene under the heading Representing the world in 3D and the subheading Photorealistic. The map can be activated by clicking the image of Montreal, Canada.

Montreal, Canada Scene

    a. This scene is described as photorealistic. What does that mean and what layers are used to produce the photorealistic map?

3. Find the High Rise Election scene under the heading Representing the world in 3D and the subheading 3D Cartographic. Activate the map by clicking the image of High Rise Election.

High Rise Election

    a. Explain how this is an example of 3D cartography.

Philadelphia Redevelopment

4. Find the Philadelphia Redevelopment map under the heading Who uses 3D cartography? The map can be activated by clicking the image Urban planners. It is the fifth map in the series.

    a. How does 3D help in the visualization of shadow produced by the proposed high-rise development?

Lesson 6-1: Understand current events in 3D
Using a virtual globe to broadcast the news

A virtual globe or Scene

The world is three dimensional, and many GIS applications require that 3D experience. Esri's Scene is a viewer that allows you to visualize data on a world sphere. The scene viewer provides a variety of tools for orientation, navigation, search, and toggling basemaps and layers.

The virtual globe of Scene is also a way to combine current events with geospatial technology. It provides a geospatial focus for a wide range of subjects and can connect many curriculum area. It provides the opportunity for learners to look at geospatial news at all levels: global, regional, and local.

Scenario

You have been asked to pick one of the scenarios below and prepare a newscast using the virtual globe as your visualization tool.

Build skills in these areas

▸ Manipulating a globe within a 3D environment

▸ Searching for geographic locations

▸ Connecting news with a world sphere

▸ Presenting material using geospatial technology

What you need

▸ Account not required

▸ Estimated time: 30 minutes

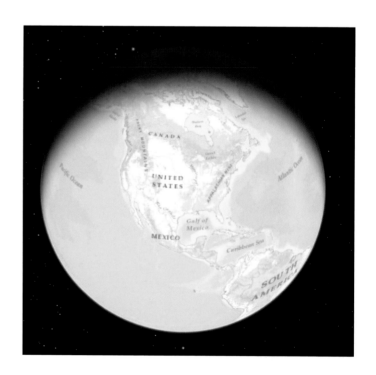

World scenario

Since 2011, Syria has been in a civil war. More than 4 million people have fled to neighboring countries, some of which are more welcoming than others. Research your topic and use the globe as your visualization.

1. Go to http://arcgis.com.
2. Click Scene on the top of the page.
3. Click the X to close the 3D Scene Viewer sign in page.
4. Take time to learn the navigational buttons:
   a. Click Home 🏠 to return to the initial position.
   b. Click + to zoom in. You can also use your mouse and scroll wheel to zoom in and zoom out, or press and hold the middle mouse button and move up or down to zoom in or out.
   c. Click Pan ✛ to pan. Click and hold the left mouse button and drag the map in the direction you want to move it. You can also pan by using the arrow keys on the keyboard.
   d. Click Rotate ↻ to rotate. Click and hold the left mouse button and drag the map in the direction you want to rotate it.
   e. Click Compass ▲ to reorient your scene north. You can also press N on your keyboard.

5. Search for Syria using the search button in the upper right corner.

6. Zoom in to the designated location.
7. Change the basemap to different maps for different information.
8. Below are some focus questions:
   a. Why is it dangerous to go through Greece without going through Turkey?
   b. What routes either by land or water would the refugees have to take to get to Germany? Name the countries and water bodies.
   c. Hungary and Austria have shown signs of resistance to the refugees and are closing or threatening to close their borders. How would this affect the routes the refugees would have to take?

Regional scenario

On January 3, 2016, approximately 100 to 150 armed men took control of the Malheur Wildlife Refuge Headquarters. Use the globe to explain why the ranchers are in conflict with the Bureau of Land Management. Research the topic and use the globe to explain. Below are some focus questions:

9. Repeat steps 1 through 7 above. Search for Malheur Wildlife Refuge Headquarters.

    • What is the history of the Harney Basin?
    • What president declared the land a preserve?
    • What is the conflict between the Fish and Wildlife Service in conjunction with the Bureau of Land Management with the ranchers?

Local scenario

In Pacifica, CA, the city has declared a state of emergency because citizens have to leave their homes due to erosion. Research the topic and use the globe to explain. Below are some focus questions:

10. Repeat steps 1 through 7 above. Search for Pacifica, CA. The specific address where the apartments are falling into the sea is 310 Esplanade Ave, Pacifica, CA.

- Is this emergency weather related? How?
- Will the residents get any financial relief?

Lesson 6-2: Visualize water landforms

Presenting the planet's geometric masterpieces

A water landform is a feature of the planet's surface that is associated with water. Water landforms are caused by geologic processes and many of them are geometric masterpieces.

Scenario

As an instructor of geography, you have decided to have your exam be an identifcation of water landforms by imagery. You are putting together a visualization of five different water landforms to be identified. The landforms are found by either longitude and latitude or by address. The water landforms to be identified are listed below:

- Gulf of Mexico Continental Shelf -88.278,27.889
- Maldives
- Kenai Fjords
- Isthmus of Panama
- Mississippi Delta -89.844,29.413

Build skills in these areas

▸ Opening a scene

▸ Searching for a location by longitude, latitude, or by address

▸ Creating slides and saving a presentation

What you need

▸ User, Publisher, or Administrator role in an ArcGIS organizational account

▸ Estimated time: 30 minutes

Open a scene

1. Sign in to your ArcGIS organizational account.
2. Click Scene on the top of the page.
3. Take time to learn the navigational buttons:
   a. Click Home 🏠 to return to the initial position.
   b. Click + to zoom in. You can also use your mouse and scroll wheel to zoom in and zoom out, or press and hold the middle mouse button and move up or down to zoom in or out.
   c. Click Pan ✛ to pan. Click and hold the left mouse button and drag the map in the direction you want to move it. You may also pan by using the arrow keys on the keyboard.
   d. Click Rotate ↻ to rotate. Click and hold the left mouse button and drag the map in the direction you want to rotate it.
   e. Click Compass ▲ to reorient your scene north. You may also press N on your keyboard.
4. Search for -88.278,27.889 using the search button in the upper right corner. A pointer appears at the location. That longitude and latitude shows the continental shelf in the Gulf of Mexico.
5. Zoom in to the location.
6. Choose an appropriate basemap by clicking Basemap. Imagery or Oceans is a good choice.

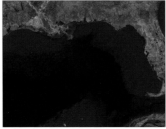

7. Click the Slides arrow.

8. Click Capture Slide.
9. Add a title.

10. Click Done.
11. Repeat steps 5 through 11, searching for the Maldives.
12. Repeat steps 5 through 11, searching for Kenai Fjords.
13. Repeat steps 5 through 11, searching for the Isthmus of Panama.
14. Repeat steps 5 through 11, searching for -89.844,29.413. That longitude and latitude shows the Mississippi Delta.
15. Click Save Scene.
16. Complete the metadata.

Lesson 6-3: Teach world time zones
Charting the hour for online students

For centuries, humans have marked time by the position of the sun. The sun rose in the east, moved across the sky, and set in the west. Solar noon occurred when the sun was directly overhead. However, solar noon in one location could occur in the middle of the night in another. What could we do to solve this problem? The answer came through establishing worldwide standard time zones. There are 24 time zones, each 15 degrees of longitude wide. The starting point for the standard time zones is the prime meridian. Travelers moving westward from the prime meridian move their clock back to earlier times (minus one hour for each time zone), while those moving eastward change to later times (plus one hour for each time zone). If a traveler goes around the world, they change not only the time but also the date. Establishing the International Date Line solved this problem. Travelers moving westward advance their calendars one day as they cross the International Date Line (Saturday to Sunday), and travelers moving east move their calendars back one day (Saturday to Friday).

Scenario
A large university that has a major department in international relations and diplomacy has asked you as a GIS educator to provide them with a lesson that will explain worldwide standard time zones. They have asked that the course be online and that it be a hands-on experience. The ArcGIS Online software is available at the university.

Build skills in these areas
▶ Opening a scene
▶ Searching and adding layers
▶ Configuring layers
▶ Saving a scene
▶ Using a scene to answer questions

What you need
▶ User, Publisher, or Administrator role in an ArcGIS organizational account
▶ Estimated time: 45 minutes

Open Scene, search for and add layers, configure, and save

1. Sign in to your ArcGIS organizational account.
2. Click Scene on the top of the page.
3. Take time to learn the navigational buttons:
   a. Click Home ♠ to return to the initial position.
   b. Click + to zoom in. You can also use your mouse and scroll wheel to zoom in and zoom out, or press and hold the middle mouse button and move up o r down to zoom in or out.
   c. Click Pan ✛ to pan. Click and hold the left mouse button and drag the map in the direction you want to move it. You may also pan by using the arrow keys on the keyboard.
   d. Click Rotate ↻ to rotate. Click and hold the left mouse button and drag the map in the direction you want to rotate it.
   e. Click Compass ⚐ to reorient your scene north. You may also press N on your keyboard.

You need to add the following layers to construct the lesson:
   - Prominent World Longitude and Latitudes
   - World Time Zones
   - Cities_tzones

4. Click + Add Layers.
5. Where it says Search for layer, type **Prominent World Longitude and Latitudes.**
6. Click +.

7. Click Done.
8. Click the pull-down arrow at the end of the layer name and choose Configure Layer.
9. Click Change Symbols and change Size (pixels) to 4.

10. Click Done.
11. Click + Add Layers.
12. Search for World Time Zones and click + World Time Zones Feature layer by Esri.
13. Click Done.
    You now have two layers on your map. You can turn these layers on and off by clicking the X in the upper right corner.

14. Think about then answer these questions:
    a. How would you describe the standard time zones.
    b. Why do some of the standard time zones have irregular boundaries on land?
    c. Explain the need for an International Date Line.
    d. How many time zones are in the continental United States?
15. Click + Add Layers again, search for city_tzones, and add it to your scene.

16. Click Done.

Notice that city_tzones seems to be floating in
    space.
17. Click the pull-down arrow to configure city_
    tzones and change the Elevation Mode to On
    the ground.
18. What cities are represented?
19. Click Done.
20. Click Save Scene.

21. Add the appropriate metadata:
    • Scene Title = World Time Zones
    • Summary = Spatial display of World Time
      Zones
    • Tags: time_zones,scene

22. Click Save Scene.
23. Go to My Content to open the World Time
    Zones scene.

Use the constructed scene to complete the chart
below:

Start City	Day/Time	Travel Direction	End City	Day/Time	# Time Zones Crossed
London	Mon 11 AM	West	Denver	Mon 4 AM	
London	Mon 11 AM	East	Denver	Mon 4 PM	
Paris	Wed 1 AM	West	Minneapolis	Tues 6 PM	
Paris	Wed 1 AM	East	Minneapolis	Tues 6 PM	
Rio de Janeiro	Friday 10 PM	West	Tokyo	Sat 10 AM	
Rio de Janeiro	Friday 10 PM	East	Tokyo	Sat 10 AM	

The ArcGIS Book, chapter 6
Questions for reading comprehension, reflection, and discussion

Teachers can use the items in this section as an assignment, an introduction, or an assessment, tailored to the sophistication of learners. Some learners can read all the sections at one time, while others are more comfortable with small segments. The questions and tasks are designed to stimulate thought and discussion.

1. The evolution of 3D mapping

    Advantages of 3D
    a. What advantage does vertical information give?

    b. What is human-style navigation?

2. Important 3D terminology: Getting the z-terminology straight

    Explain these terms (including the difference between them where appropriate):
    a. Maps and scenes
    b. Local and global
    c. Surfaces
    d. Real size and screen size

Representing the world in 3D
e. Define photorealistic
f. What makes 3D Cartography powerful?
g. What two factors are involved to create a feeling of virtual reality?

What makes a great scene?
h. What is meant by implying a 3D scene is designed to be immersive?
i. What are the three choices of styling 3D content?
j. List two ways to illustrate thematic views.

3. Thought Leader: Nathan Shephard: The rise of the 3D cartographic scene

Nathan talks about the benefits of communicating spatial data in 3D and the fact that cartographers are no longer limited to two dimensions.

4. Who uses 3D cartography?

Go to http://arcg.is/1RrSZT7 and investigate three of the maps.

More online lessons
At the end of chapter 6 in the online version of *The ArcGIS Book* (http://learn.arcgis.com/en/arcgis-book/chapter6/), you'll find a Learn ArcGIS lesson to analyze flooding in Venice, Italy, both in 2D and 3D.

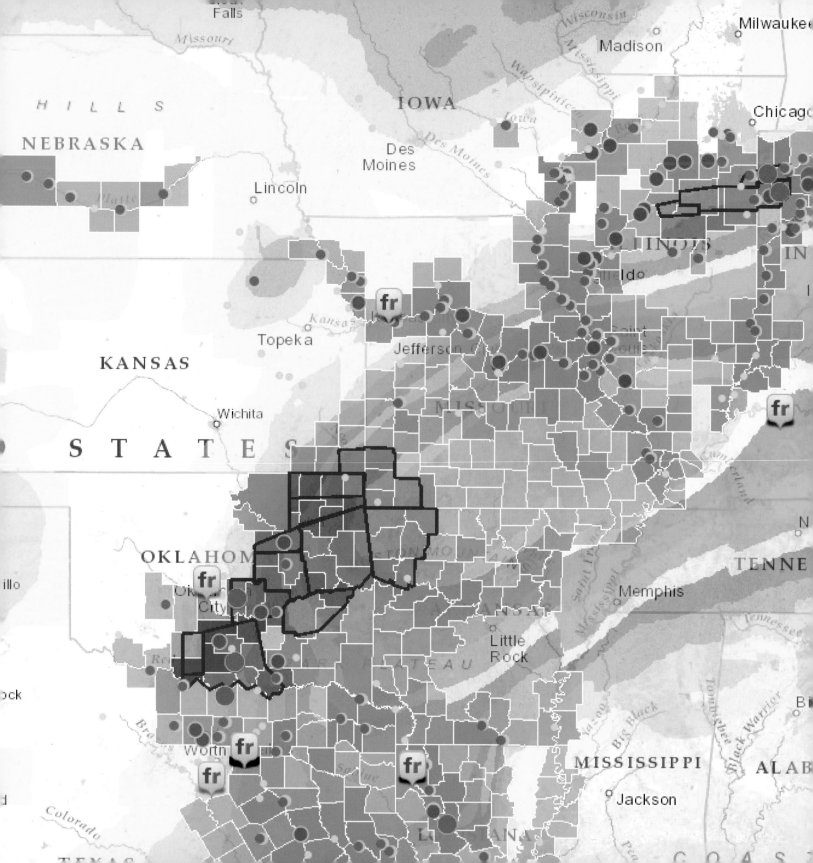

The Power of Apps
Focus and individualize your work with online applications

"With billions of users worldwide, apps are a technology trend that has captured the world's attention. Online maps provide the information that powers the use of GIS. And every map has an interface—a user experience for putting that map to use. These experiences are apps, and they bring GIS to life for users." (*The ArcGIS Book,* p. 92)

Have you ever been out to dinner and noticed that no one is talking to you? They are on their smartphones, their iPads, or their Androids. It is the invasion of the apps. You can find a place to eat, a hotel, a ball game; sitting down, standing, or walking, you can look up almost any information. Wherever you are you can check the weather forecast or know when a plane is landing.

Apps are lightweight computer programs designed to run on the web, smartphones, tablets, and other mobile devices. While all such applications serve a purpose, GIS apps are even more useful because they are mapcentric and spatially aware. GIS apps can collect data, alert you to geographic events, and answer questions through analysis. In this chapter you will:

- Visit ArcGIS Marketplace and pick out two apps of interest to you, then download and investigate them.
- Participate in a collaborative data experience showing participant skill and location of everyone using the *Instructional Guide for The ArcGIS Book.*
- Be introduced to a flood modeling app, an Explorer app, and a geo-alert app.
- Begin exploration of a topographic app.
- Continue advanced exploration of a topographic app.
- Download and use Snap2Map.

The lessons in this chapter offer practice (and entertainment) in downloading and experiencing apps. Use the questions at the end of this chapter to support your reading comprehension, reflection, and discussion of the narratives presented in the corresponding chapter 7 of *The ArcGIS Book.*

Introductory activities

Video
Go to www.youtube.com and search for the video by title, "A map for every story."

A map for every story
Unique stories are being created and shared using ArcGIS Online from Esri. Explore the world of web maps, create your own, and discover stories with a cloud-based, collaborative system.

Activity
Investigate ArcGIS Marketplace
marketplace.arcgis.com

1. What is marketplace?
2. Who can use the apps?
3. Select two apps and explain why you would want to use them.
4. Download them to your mobile device and enjoy.

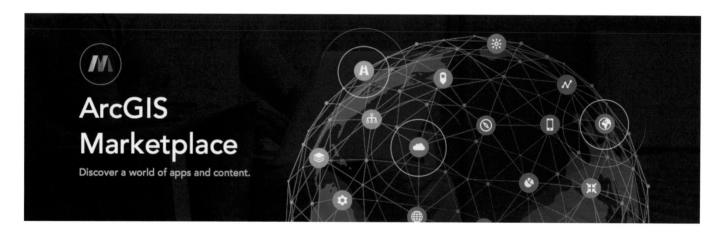

The first lesson in this chapter introduces you as a user to a collaborative data collection. The other lessons in this chapter are designed to show you how to download apps and get a taste of the app experience. You will either download the apps to your mobile device or access them online and then perform a series of activities. The lessons are short and teachers can use them as drill-for-skill, fillers, informationally, and for fun for their students.

Lesson 7-1: Collect data collaboratively
Learning from *The ArcGIS Book*

As *The ArcGIS Book* says, apps provide the interface for the efficient collection of spatial data so you can leverage your handheld device as a data collection tool. As a user of *Instructional Guide for The ArcGIS Book,* you are being asked to enter your location and learning environment into the cloud using an editable feature service (a shared tool) and working on the map collaboratively. Record your location by clicking on the map and recording the following information:
* Higher Education
* K-12 Education
* Online Program

Scenario
The ArcGIS Book team wants to collect data about the learners using *The ArcGIS Book.*

Build skills in these areas
▸ Opening a map with an editable feature service.

▸ Contributing information to a collaborative map.

What you need
▸ Account not required

▸ Estimated time: 5 minutes

1. Go to: http://arcg.is/1qKlgvQ
2. Click Edit on the top of the page.
3. Click Learners in the table of contents.
4. Click to add a point on your location on the map.
5. Fill out the information requested.

You are now part of an online collaborative map collection.

Lesson 7-2: Assess risk of inundation
Using STORMTOOLS

There are a number of interactive online apps that explore the potential impact of sea level rise (SLR) on the US coast – Climate Central's Surging Seas, the PBS site called Will Your City Be Under? There's a Map for That, the NOAA Sea Level Rise Viewer, and STORMTOOLS, to name just a few. As the Ocean State, Rhode Island faces serious challenges from sea level rise in the coming years. STORMTOOLS is a web-based app created specifically for residents of Rhode Island to better understand their risk from coastal inundation. The STORMTOOLS app is part of the Shoreline Area Special Management Plan (SAMP) of the RI Coastal Resources Management Council (RICRMC).

Scenario
You would like to buy property on or near the Rhode Island coast to build a family destination resort. You have narrowed down the possibilities to properties in three Rhode Island towns: Bristol, Middletown, and Narragansett. Use the STORMTOOLS app to determine the risk of inundation to each of these properties and make a recommendation about which one represents the best choice.

Build skills in these areas
▸ Opening the STORMTOOLS app
▸ Creating map notes layers at three specified locations on The STORMTOOLS map

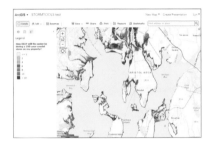

▸ Comparing the risk of inundation at each of three specified sites
▸ Identifying the best option for project construction

What you need
▸ Account not required
▸ Estimated time: 45 minutes

Open the STORMTOOLS app.
1. Go to http://arcgis.com.
2. Search for stormtools.
3. Select STORMTOOLS for Beginners from the search results.

Create map notes layers at three specified locations on the STORMTOOLS map

The three properties under consideration are located at the following coordinates:

Property	Longitude	Latitude
A Narragansett	-71.430	41.454
B Middletown	-71.241	41.506
C Bristol	-71.267	41.651

1. Turn off the future Sea Level Rise layer.
2. Locate Property A on the map by entering its coordinates in the STORMTOOLS Find window as follows: **-71.431, 41.454**. Click the search tool.

3. When the Search Result window appears, click Add to Map Notes.

4. Click Modify Map above the map.
5. Click the map note to open the popup and click Edit.
6. Click Change Symbol and choose the A-Z symbol set.
7. Choose A and make the Symbol size 30.

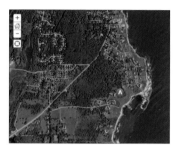

8. Repeat the process to locate Properties B and C.
9. Click Edit above the table of contents to end the editing process.
10. Save your map as Flood Risk Comparison. When finished, your map should show all three properties under consideration as shown below.

Compare the risk of inundation at each of three specified sites

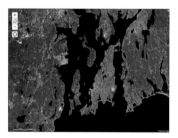

For the purpose of estimating property risk, assume that the proposed resort will occupy significant space surrounding the location of the map note.

1. Zoom to each of the three properties, one at a time.

2. One by one, turn on the layers related to the risk of inundation and briefly describe the risk to the property for each scenario.
   Notice that you can view sublayers reflecting different scenarios under several of the layers.

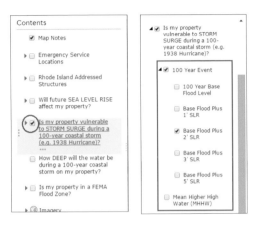

3. Consider the answers to each of these questions as you compare Properties A, B, and C.

   a. Will future Sea Level Rise affect my property?
   b. Is my property vulnerable to Storm surge during a 100-year coastal storm (For example, 1938 Hurricane)?
   c. How deep will the water be during a 100-year coastal storm on my property?
   d. Is my property in a FEMA Flood Zone?
   e. Based on your analysis with the STORMTOOLS data, identify the property that seems to be the least vulnerable to flooding from Sea Level Rise and/or storm events. Which one seems the most vulnerable? Which one would you like to purchase? Why?

Lesson 7-3: Get Answers with an app
Solving problems with Explorer for ArcGIS

In Explorer you can discover and visualize even without an organizational account. You can find places and use a broad range of Esri map collections and basemaps. Use the Explorer for ArcGIS app to formulate an answer.

Explorer for ArcGIS download from iTunes
https://itunes.apple.com/us/app/explorer-for-arcgis/id860708788

Explorer for ArcGIS download from GooglePlay
https://play.google.com/store/apps/details?id=com.esri.explorer

Scenario 1
An ecoregion is a relatively large unit of land defined by its ecology or environmental conditions, such as climate, landforms, and soil characteristics. Ecoregions boundaries correspond well with land cover. You have been asked to identify and explain the ecoregion and the prevalent land cover for the following geographic areas:
- Your specific location
- Specific political boundary such as a county or state
- A natural boundary such as a watershed

Build skills in these areas

▸ Downloading an app

▸ Navigating the app

▸ Using the app to solve problems

What you need

▸ Account not required

▸ Estimated time: 45 minutes

Use the Explorer app
1. Double-click the app on your mobile device.
2. Click Continue without signing in.
3. Click Find Maps.
4. Search for Ecological Divisions.
5. Click Ecological Divisions.
6. Search for the following geographic locations:
   a. Your specific location.
   b. Specific political boundary such as a county or state.
   c. A natural boundary such as a watershed.
7. Under Featured Content click Collections.
8. Click Landscape Maps.
9. Click USA Land Cover 2011.

Questions for Scenario 1
1. Define land cover.
2. Search for NLCD legend and explain the colors.

3. Write an analysis comparing the ecoregion of the specific geographic location to its Land cover.

Scenario 2
Of course, forest fires most often occur in remote areas that are hard to measure. As the team leader in the field, you have just received satellite imagery of the Rim Fire in Yosemite Park on your phone. You have been asked to give a rough estimate of the square-miles area of the fire so that the administrative office can plan a strategy.

1. Double-click the app on your mobile device.
2. Click Continue without signing in.
3. Click Find maps.
4. Search for Fire Rim.
5. Click Rim Fire at Yosemite.
6. Your task is to find the area of the fire in square miles:
   a. Click the Measure area at the bottom.
   b. Digitize around the fire.
   c. Record the square miles.
7. Repeat steps 4 through 6 but search for Glacier National Park Fires.
8. Find the area of both Glacier Fires.

Scenario 3

You are fishing in the Gulf of Mexico and have landed in a school of red snapper. You are catching so many fish that you want to send the locational information (latitude, and longitude) to your companion boat.

1. Double-click the app on your mobile device.
2. Click Continue without signing in.
3. Change the basemap to Imagery or Oceans.
4. Tap and hold a location.
5. Drop a pin.
6. Record the longitude and latitude.

Lesson 7-4: Investigate relationships
Alerts and news from the QuakeFeed Earthquake app

The QuakeFeed Earthquake Map, Alerts and News provides numerous tools for learning. Earthquakes are tracked in real time and they can be filtered by magnitude or distance from a location. This app gives you a layer of classified plate lines and eight different basemaps.

Scenario

As a high school Earth Science teacher, you have decided to engage your students in the subject matter through their obsession with their phones. You would like for them to use the QuakeFeed Earthquake app to investigate the relationship between the magnitude of earthquakes and the type of plate boundaries.

Build skills in these areas

▸ Downloading an app to a mobile device

▸ Manipulating and using an app

▸ Using the mobile device to answer a question

What you need

▸ Account not required

▸ Estimated time: 30 minutes

1. Download the app. QuakeFeed Earthquake Map, Alerts and News.
2. Click the app to open it.
3. At the bottom of the screen, choose Filter and select a magnitude of 5.0 and higher.

   a. What types of plate lines are located where the earthquakes are high?
   b. Describe the two areas on the map that have the most seismic activity.

Lesson 7-5: Exhibit changes over time
The USGS Historical Topographic Map Explorer app

In 1879, the USGS began to map the US topography and continued to do so until 2006. As the years passed, making good use of satellite imagery, the USGS continuously updated map versions of each area. USGS topographic maps were provided at a variety of scales ranging from 1:24,000 to 1:250,000. Today they are available digitally to view and download via the following:

(TopoView app)
http://ngmdb.usgs.gov/maps/TopoView/viewer/#4/40.00/-100.00

The most current maps are available from The National Map
http://nationalmap.gov/

USGS Historical Topographic Map Explorer, a web-based app.
http://historicalmaps.arcgis.com/usgs/

In this lesson, you will learn how to use one of those apps, the USGS Historical Topographic Map Explorer (at the link above) to find historical maps that reveal and portray changes over time at your location (for example, changes in land use, roads, neighboring structures). You will use the USGS Historical Topographic Map Explorer App to identify maps you would like to use in your project.

Scenario
Your class has decided to create a museum exhibit about the history of your school, the Lanier Middle School in Fairfax, Virginia. Separate teams will design different components of the exhibit, and you are part of a team that will focus on changes over time in landscape, land use, and settlement patterns.

Build skills in these areas
▸ Opening the USGS Historical Topographic Map Explorer
▸ Searching for historical topographic maps of a specific location
▸ Identifying maps that support your project goals

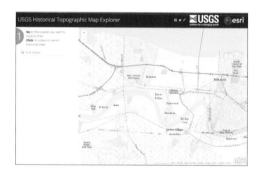

What you need
▸ A computer or mobile device with Internet connectivity
▸ Estimated time: 30 minutes

Open the app and search

Open the USGS Historical Topographic Map Explorer and search for historical topographic maps of a specific location.

1. Go to the app at http://historicalmaps.arcgis.com/usgs/ .

2. In the search box (Find a Place), enter the following address:

   3801 Jermantown Road, Fairfax, VA 22030

3. Select it from the dropdown list (it will be first).

4. Zoom out just slightly so you can see the Lanier Middle School Building.

5. Click on it.

   A red + appears where you clicked and a list of maps that include that location are displayed across the bottom of the screen.

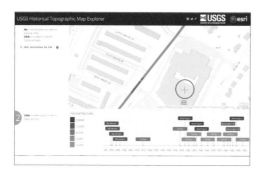

Note that a color key to the various map scales is displayed on the left. For our web app, we want maps at the 24,000 or 62,500 scales because they will provide the most detail.

Identify maps that support your project goals

1. Point to the earliest 62,500-scale map and you'll see a thumbnail of the map (Fairfax, 1915).

2. Click the timeline marker to display the map in the main window.

   The 1915 map of Fairfax is displayed and listed

in the left map panel. You can remove the map by clicking the x in the upper left corner. You can turn it on and off by making it transparent. Notice that the red + still indicates the location of the school even though the school is not on this map. You may want to zoom out slightly to get a sense of the area at this scale.

3. Choose another map from the list below the display.

   The next map chronologically is Fairfax 1944. When you add it to your map your display looks like this.

   a. Toggle between the two maps by making the upper one transparent.
   b. Zoom out to get a sense of the larger surrounding region.
   c. Make note of changes that occurred in this 30-year period as reflected in differences between the maps.

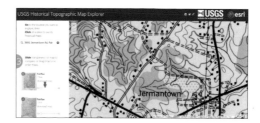

4. Continue to add and explore more maps to your display: 1955, 1966, 1979, and 1994.

    When you've added all the maps, your contents bar looks like this.

5. Explore and compare the maps to identify ones that tell the clearest story of changes around the Lanier Middle School from 1915 to the present. Note that you can download any of these maps as a geopdf. The download link is next to the map thumbnail in the table of contents.

    If you have an organizational account, you can add these maps to ArcGIS Online.

Lesson 7-6: Add history to ArcGIS Online
Topographic maps from the Living Atlas of the World

In this section you will add historical topographic maps to an ArcGIS Online map from the Living Atlas of the World.

Build skills in these areas

▸ Adding historical topographic maps as layers to ArcGIS Online

▸ Creating a web application to show change over time at one location

▸ **What you need**

▸ An ArcGIS Online organization account

▸ Estimated time: 1 hour

Scenario

Your team has decided to create a web application for the exhibit that will enable viewers to compare historical topographic maps of the school's surroundings at different historical dates. Your responsibility is to create the web application. Others on your team will provide content based on research.

Create a new map in your ArcGIS Online account

1. Open a new browser tab and go to ArcGIS Online. Sign into your organizational account.
2. Open a New Map
3. After using the find address box to zoom to the Lanier Middle School (3801 Jermantown Road, Fairfax, VA 22030), click Add to Map Notes to place a marker there.

4. Click the map note.
5. Click Edit.
6. Rename the map note **Lanier Middle School**.
7. Click Edit to stop editing.
8. Also rename the map notes layer in the table of contents, **Lanier Middle School**.

9. Save your map as **Lanier Middle School**.
10. Fill in the map information with tags and a summary (**Historical topo maps of the Lanier MS area**).

Add historical topographic maps as layers to an ArcGIS Online map

1. Click Add.
2. Click Browse Living Atlas Layers.
   a. Under All Categories, choose Historic Maps. You will need to select maps from both the 1:62,500 scale and the 1:24,000 scale groups.
3. First add the 1:62,500 group.

4. Click the more options symbol (three dots).
5. Choose Image Display Order from the dropdown list, and in the window that opens:
   a. Leave An attribute at Date Current.
   b. Change the Highest priority value to 1915. You are adding the 1915 map from the 1:62,500-scale group.
   c. Click Apply.
   d. Click Close.
6. Click more options again and click Copy.

A copy of the USA Historical Topo Maps 1:62,500 has now been added to your map.

7. Go to Image Display Order on this new copy of the layer and change 1915 to 1944. You can see the two map layers that you've added by toggling on and off.
8. Rename the layers in the table of contents with the appropriate year:
   a. For the first map you added (the lower one), click more options and click Rename.
   b. Change the layer name to 1915.
   c. Repeat this with the layer copy.
   d. Rename this layer to1944.
9. Return to the Living Atlas Historical Maps group (Add>Browse Living Atlas Layers>Historical Maps).
10. Add the USA Historical Topo Maps 1:24,000.

11. Use the same procedure you did with the first group of maps, to add maps from the 1:24,000-scale group. Go to Image Display Order in that layer and replace 1800 with 1955 (the next map you want to add).
12. Click Apply.
13. Click Close.
14. Copy the USA Historical Topo Maps 1:24,000 layer 3 times because you want to add 3 more maps from this group (1966, 1979, and 1994).
15. Repeat the selection procedure to add each of these three layers to your map.

Save your map frequently.

16. Rename the layers appropriately: 1955, 1966, 1979, and 1994.

When you have added all the layers, your map will look like this.

The final layer in your map will be a 2015 topographic map.

17. Go to Add>Search for layers>In ArcGIS Online and search for usa topographic.
    a. In the Search results, choose and add USA Topo Maps.

18. For each of your map layers, click More Options (elipsis) and Hide in legend.

Create a web application to show change over time at one location

1. Click Share.
2. Click Create a New Web App.
3. On the Create a New Web App page, leave the selection to Show All and scroll down until you see the Map Tools app. Choose it, and click Create App.

4. Add a map summary (Historical topo maps of the Lanier MS area) and any additional tags you want to the information details for your app.
5. Click Done.
6. In the map configuration window, check off all your map layers.
   You need to do this if you want to share the map publically because it contains restricted content.
7. Click Save and click View.
8. Change settings in Configure Web App.
   a. General: leave default settings.
   b. Theme: choose colors for map components (you can change these later).
   c. Options: check Home Button, Zoom Slider, and Layer List. Uncheck all others.
   d. Print: uncheck Print Tool and display all Layout Options.
   e. Search: uncheck Enable search tool.

9. Click Save.
10. From your My Content list, open the web app you have created for the museum exhibit.

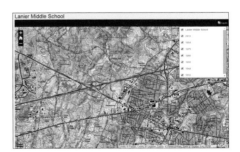

Lesson 7-7: Profile your community
Creating a Story Map Tour with Snap2Map

One of the most popular apps in recent years is Snap2Map. Using only your smartphone and this free Esri app, you can create a Story Map Tour in minutes, and publish it directly to your ArcGIS Online account from your smartphone. In no time at all, your story map is ready to be viewed on a smartphone, tablet, or laptop. Snap2Map is available as a native app for iOS and Android devices such as smartphones and tablets. To use Snap2Map all you need is publisher or administrator permissions in your organization.

Scenario

You work for a large real estate agency and your office has properties for sale in a number of communities across the region. Prospective buyers from outside the area frequently ask, "What's it like there?" when presented with a potential property in a particular community. The owner of the agency has decided that the office needs some simple Story Map Tours that will give prospective buyers a visual overview of the communities where these houses are for sale. You have been asked to install Snap2Map on your smartphone and use it to create a Story Map Tour called What's it Like in [name of your town].

Build skills in these areas

▸ Downloading an app to a mobile device

▸ Manipulating and using an app

▸ Using the mobile device to create a Story Map Tour

▸ Publishing the Snap2Map tour to your ArcGIS organization

What you need

▸ ArcGIS Online organization account

▸ Snap2Map app

▸ Mobile device

▸ Estimated time: 30 minutes

Download an app to a mobile device

1. Download and install Snap2Map onto your Android or iOS smartphone.
2. Open the app when the installation is complete.
3. Click Start.
4. Sign in to your ArcGIS organization.

Create a Story Map Tour

1. Click Create New Map Tour.
2. Give your tour a title and description and choose an appropriate basemap.
3. Click Next.

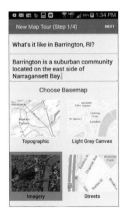

The next step is to add pictures to your tour. Note that on an iOS smartphone, you can take new pictures or choose images already stored on the device. On an Android device, you must choose photos already taken and stored on the phone. If you haven't done so already, use your smartphone to take pictures of your community that would give newcomers a sense of what it's like there.

4. Choose the folder where your pictures are stored and select the photos you want to use in your tour. You can choose them one-by-one or all at once.
5. Add a title and description for each photo. (Keep in mind, you can also add or edit these later on your computer.)
6. If you see a notice that says "Photo not geo-tagged, Click to Add," click the picture. A map appears, and you can choose where your photo belongs by zooming and panning the map until the pin is in the correct location.
7. When you're satisfied with the location of the pin, click Done.

Publish the Snap2Map tour to your ArcGIS organization

After clicking Done, you will be given the option to save your tour as a draft, cancel it, or publish it. If you save it as a draft, you can come back to it later to add more photos, change descriptions, and so on. If you choose Publish, your tour will be published to your ArcGIS organization. You can make edits to the tour from your computer as well.

1. Click Publish.

2. When the map has finished publishing, open it up in your Internet browser, or click Done if you prefer to explore later.
3. To edit the map on your computer, go to www. arcgis.com and sign into your account.
4. Open your Story Map Tour from the My Content list where it has been saved.
5. To edit the map, click the gray Switch to builder mode button in the upper right.
    At this point, you can change the basemap, add more pictures, change descriptions, modify the color scheme, and share your map.

6. Be sure to share your Story Map Tour with everyone so prospective buyers can open it.

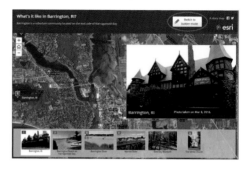

More online lessons

At the end of chapter 7 in the online version of *The ArcGIS Book* (http://learn.arcgis.com/en/arcgis-book/chapter7/), you'll find a Learn ArcGIS lesson to use Web App Builder to create an Oso mudslide swipe map app.

The ArcGIS Book, chapter 7
Questions for reading comprehension, reflection, and discussion

Teachers can use the items in this section as an assignment, an introduction, or an assessment tailored to the sophistication of learners. Some learners can read all the sections at one time, while others are more comfortable with small segments. The questions and tasks are designed to stimulate thought and discussion.

1. The rise of spatially intelligent apps

   a. What is an app?

   b. How is a GIS app unique?

2. Thought Leader: Abhi Nemani: Shaking things up in the City of L.A.

   a. What does the notion of government as a platform mean?

3. Case study: US Geological Survey
   In 2009, the US Geological Survey began the release of a new generation of topographic maps in electronic form, and in 2001, complemented them with the release of high-resolution scans of historical topographic maps of the United States dating back to 1882. View these using the USGS Historical Topographic Map Explorer.

4. Where do apps come from?
   Solve a problem with an app

   a. Apps come from the following places:

   b. Solve a problem with an app in the following ways:

5. Quickstart: Use out-of-the-box apps, build apps without having to write any code, or code your own apps from scratch

   a. What is the "app revolution" and why does it represent a major change in how we implement GIS?

Your Mobile GIS
Collect data in the field with your smartphone or tablet

Does the data you collect need to be mapped? Not necessarily. If the data is not informed by or associated with a location, GIS is not the tool to use. Nonlocational data can be recorded in a table and sorted. Much data, however, is most useful when in geographical context. If you intend to map your data, then it must have a spatial component. How do we record spatial data?

Historically, we went into the field with a paper and a pencil and maybe a GPS and we recorded data. We made a mark on a map, wrote down the latitude and longitude, and mapped the data when we came back. Now we can do all of this in the field: Individual collectors can work with their own devices on the same collaborative map. You can stand with your mobile device and see points from other collectors displayed. You can even set up your collection device so you don't have to be online with a phone service or Wi-Fi connection.

However, before you start collecting data there is work to be done. First, you need to plan your data collection. What information do you need? What type of data is the information? Remember, if you are going into the field to collect, you want to get all the data at one time.

Second, you need an editable feature service or web layer to push into the cloud so everyone can access the same layer and work collaboratively. An editable feature service/editable web layer allows you to save features over the Internet, display symbology of the feature and, most important, allow collectors to perform edits.

Third, you need to choose from a variety of collection apps that can be downloaded to your mobile device. Use the questions at the end of the chapter to support your reading comprehension, reflection, and discussion of the narratives presented in the corresponding chapter 8 of the *The ArcGIS Book*.

Introductory activities

Video
Go to video.esri.com and search for each video by the title.

Getting Started with Collector for ArcGIS

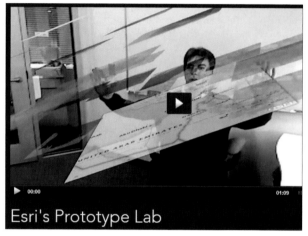

Esri's Prototype Lab

Getting Started with Collector for ArcGIS (2013): *Collect and update spatial field data and plan routes, and get directions.*

Esri'sPrototype Lab (2014): *Get a look into Esri's future developments.*

Activity

Spatial versus nonspatial data

GIS is not the tool to use if the data that you are collecting does not have a spatial component. For example, if you are recording the height of a group of people, a tool like Microsoft Excel would be better.

Spatial data has a locational or geographic component. It is referenced to a location on the surface of the earth.

Nonspatial data does not have spatial or locational qualities.

Below is a list of data. Divide the data into spatial and nonspatial categories.

person's height	weather	rivers
elevation	recipes	phone numbers
book titles	television shows	cities

Spatial:

Nonspatial:

Lesson 8-1: Assess collected information
The tree inventory map

With mobile devices, you can gather information about types of trees, their height, and the canopy diameter. Such data can be helpful in many ways. For example, you could efficiently harvest trees for lumber while responsibly including in your planning the conservation and protection of the ecosystem.

Scenario

A group of teachers at the annual Esri Teachers Training Teachers Institute collected information about trees. You are trying to duplicate this experience for a local institute you are running. Your task is to investigate the feature service.

Build skills in these areas

▸ Investigating attributes

▸ Speculating on analytical functions

What you need

▸ Account not required

▸ Estimated time: 20 minutes

Find the Tree inventory map, for use in this lesson, in chapter 8 of the thearcgisbook.com, located under the heading What is Collector for ArcGIS?.

Tree inventory

With mobile devices, tree inspectors working in suburban neighborhoods gather information about tree health and maintenance work.

1. Click Tree inventory.
2. Click T3g2013_Redlands_trees, click Open, and Add layer to map.

3. Zoom in on the tree icon located near Los Angeles.

4. What can you tell about the spatial distribution of the trees?
5. Open the table and write down the attributes that were collected.

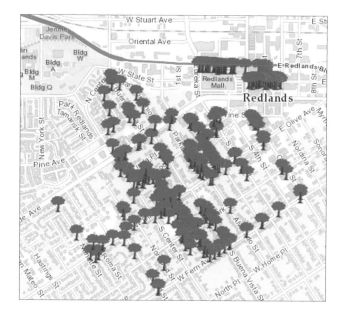

6. How could you analyze this data?

7. Is there an outlier? What were the initials of the collector?

8. Were there any attachments?

Lesson 8-2: Own a feature service

Accessing a service definition

This lesson introduces you to service definitions. A service definition is a file that is produced every time a feature service is created. It contains all pertinent information about the feature service. It allows you to duplicate a preexisting feature service, add the service to your organization, and transfer all ownership to you.

Scenario

After investigating the T3G2013_Redlands_trees collection, you have decided that this is just the template you need to have a group of instructors use in a training session you are providing. However, the T3G2013_Redlands_trees collection feature service is not editable. You have to see if you can get access to the service definition, upload it to your organizational account, and publish it. Then you will have ownership of the feature service. Ownership means that you and only users that you share with can collect data.

Build skills in these areas

▸ Finding a service definition template

▸ Downloading a service definition

▸ Publishing as a unique feature service with ownership

▸ Collecting data with the feature service

What you need

▸ Publisher or Administrator role in an ArcGIS organizational account

▸ Estimated time: 30 minutes

1. Go to arcgis.com
2. Click Sign In.
3. Click Search.
4. Search for LearnResource.

5. Download the file Tree Collection by clicking Open and save to your computer.

6. Sign in to your ArcGIS organizational account.
7. Click My Content on the top of the page.
8. Click Add Item From my computer.

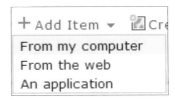

9. Browse to the Tree-Collection.sd file and choose file.
10. Add the appropriate tags.
11. Check Publish this file as a layer.

12. Click Add Item.

The creating service icon is an indication that the service definition is now being published with its own service definition and in your account and under your ownership.

Watch the Creating Service icon build the feature service and the service definition.

13. After the feature service is created, click Open and click Add layer to map.
After the feature service is added to the map, you will see the Edit icon on the top of the page.

14. Click Edit.
When you click the Edit icon, the trees collector icon is exposed.
15. Click the tree icon.
16. Click the location of the tree on the map. The collection menu is exposed.

Located at the bottom of the collection menu, Browse allows you to upload attachments from your computer. Once the attachments are loaded to the cloud, they become part of the map. You are now the owner of the editable feature service.

Below are sites that have additional service definitions that can be downloaded.

Esri Collector Templates
http://www.arcgis.com/home/group.html?owner=esri_collector&title=Collector%20Templates

James Madison University Geospatial Semester
http://learngis.maps.arcgis.com/home/group.html?id=a6cc2e7ff5c14aa2a6b806b2c042a5e1

Lesson 8-3: Map tree data
Using Collector for ArcGIS

Collector for ArcGIS allows you to use your smartphone or tablet to collect data.

Scenario

You have been assigned the task of preparing a map using the tree collection service you prepared in lesson 8-2. Remember, this is the collection template that you need to have a group of instructors use in a training session. You now have it uploaded, shared, and ready to access on your mobile app.

Build skills in these areas

▸ Downloading the Collector for ArcGIS app to mobile device

▸ Creating a group with an editable feature service

▸ Sharing a map with workers in the field

▸ Collecting data

What you need

▸ Publisher or Administrator role in an ArcGIS organizational account

▸ Estimated time: 2 hours

Create a map and share with a group in ArcGIS Online

1. Log in to your organizational account.
2. In My Content, find the Redlands_tree feature service and the Redlands_tree service definition.

3. Add the Redlands_tree feature layer to a new map.
4. Save the map with the appropriate metadata.

5. Under Home, select Groups.
6. Click Create a Group.
7. Name the group Tree_Collection.
8. Write a summary for the group: Content for Tree Collection.
9. Write a description for the group: **This survey collects data about collector's name, the type of tree, height of tree, diameter of tree, comments, and attachments.**
10. Insert the appropriate tags.

Status:
○ Private (Users cannot find this group and can join by invitation only.)
⦿ Organization (Users within your organization can search for and find this group.)
○ Public (Users can search for and find this group.)

11. Under Status, click Organization.
12. If you have a .jpg image that you want to represent the group, click the icon button and add the image.
13. Return to your map in My Content and share with the group you have just created.

Your map is now ready to be viewed in Collector. Collector will not recognize a map unless it is shared with a group.

Download Collector for ArcGIS

To complete this lesson, you will be asked to download and learn to navigate the mobile component of GIS: Collector for ArcGIS. Collector for ArcGIS is designed to improve the efficiency of field work and the accuracy of your GIS. Use your mobile device to collect and update information in the field, whether connected or disconnected.

1. On your mobile device, go to the app store, search for Collector for ArcGIS, and download the app.
2. Open the mobile app and sign in to your organizational account.

3. Click ArcGIS Online.

Accessing map, navigating and collecting on Collector

1. Open the Collector app and you will see the Tree Collection map you made and shared with the group; it's aptly named Tree_Collection.
2. Open the map and allow Collector to access your location using the app. The Collector device has different icons on iOS devices and Android devices. However, all of them universally do the same thing.
3. Familiarize yourself with the Collector app. Whichever operating system you are using, figure out how to do the following:
   - Change the basemaps.
   - Measure.
   - See a legend.
   - Collect a point.

The point collection is critical. As a general rule, to collect a point you need to do the following regardless of the operating system you use:
   - Click the GPS point.
   - Fill out the collecting form.
   - Submit the form.

The points you and the other collectors are collaboratively gathering on the map are saved in the cloud.

Lesson 8-4: Collect data offline
Syncing it later

Sometimes field collection involves being in an area where you have no Internet connection. It is possible to prepare and upload your data, collect data in a disconnected environment, and sync the data when a connection is available.

Scenario

By necessity, gathering information about trees is often done in the way you're about to do it. You are in an area where no phone or Wi-Fi connection is available. In such a case, you need to prepare maps, upload, collect in a disconnected environment, and sync at a later time in a connected environment.

Build skills in these areas

▸ Preparing and uploading data

▸ Collecting data while disconnected

▸ Syncing data when your connection is reestablished

What you need

▸ Publisher or Administrator role in an ArcGIS organizational account

▸ Collector for ArcGIS

▸ Estimated time: 30 minutes

Open the Redlands_trees editable web layer from the group Tree_Collection

1. Log in to your organizational account.
2. In My Content, locate the Redlands_trees feature layer.
3. Click the Redlands_trees feature layer.
4. Click Edit.
5. Scroll to the bottom of the page and check Enable editing and allow editors to Add, update, and delete features.
6. Check Enable Sync (disconnected editing with synchronization).

Editing	☑ Enable editing and allow editors to:
	◉ Add, update, and delete features
	○ Update feature attributes only
	○ Add features only
Export Data	☐ Allow others to export to different formats.
Sync	☑ Enable Sync (disconnected editing with synchronization).

7. Click Save.
8. Add the Redlands_trees layer to a new map.
9. Zoom to a point of interest.
10. Save the map with the appropriate metadata.
11. In My Content, click the map you just saved.
12. Click Edit.
13. Check Enable offline mode.

Delete Protection	☐ Prevent this item from being accidentally deleted.
Save As	☑ Allow others to save a copy of this item.
Offline Mode	☑ Enable offline mode.

14. Share the map with everyone and with the Tree_Collection group you previously made.
15. Click Done.

Use Collector for ArcGIS, download, and sync

1. On your mobile device, click the Collector for ArcGIS app.
2. Click ArcGIS Online.
3. Sign in to your organizational account.
4. Find the map you just created. Notice the cloud download button.
5. Click the cloud.
   In the scenario you have been given, you are asked to collect data in a remote park.
6. Zoom in and choose your work area, a park of your choice.
   Choose a very small work area, keeping in mind that you have to download that part of the basemap.
7. Click Map Detail.
8. Click Download.
   You will see the small map area you selected being downloaded to your mobile device.
9. Collect your points by the GPS locator or by sight on the image.
10. Click the map and click Upload sync.
11. This button syncs all the points that have been collected into the one collaborative map in the cloud.

Lesson 8-5: Enable citizen collaboration

Creating an editable web layer to map graffiti: Survey123

Two uses for mapping graffiti come immediately to mind: the first use is to classify the type of graffiti that is present and the second is to specify what graffiti needs to be removed by the municipal authorities. Since graffiti has a spatial component, clusters can be found across the city representing gangs, political, or religious trends.

Scenario

As a GIS consultant you have been asked to create an editable web layer that is accessible online to the general public. This editable web layer will enable citizens to work collaboratively on one map. The size of the group is negligible so many citizens can work at one time. A map containing the editable web layer can be made anywhere in the world and be accessed wherever there is an Internet connection. The editable web layer will be configured with attribute domains to enforce data integrity. A successful editable web layer with domains requires proper planning and implementation.

Build skills in these areas

▸ Downloading Survey123 Connect

▸ Creating a Survey123 form

▸ Creating a domain

▸ Using a thumbnail

▸ Publishing a XLS Form

▸ Sharing your survey

▸ Setting a survey title

▸ Using images in your survey

▸ Creating subdomains

What you need

▸ Publisher or Administrator role in an ArcGIS organizational account

▸ Estimated time: 2 hours

Open the Survey123 software on your computer

1. Go to http://survey123.arcgis.com/#/.
   You must have an ArcGIS Online account. You can get a 60-day free trial.
2. If you don't have one, click Try It Now and sign up for an organizational account then sign in.
3. Click Survey123 Connect.
4. Download Survey123 Connect for Your Operating System (Windows, Macintosh, or Linux).
5. Save the file.
6. Run the executable.
7. Double-click to open the software.

Construct your survey with data domains for integrity of data

1. Click New Survey.

2. Name the survey **Graffiti.**
3. Choose Template.
4. Click OK.

You are presented with a formatted Excel spreadsheet, so take a few moments to orient yourself to the spreadsheet. There are three main headers at the top.

- Type is the supported form type.
- Name is the Database field name and it should be unique and contain no spaces.
- Label is the question that appears in your survey.

You will design your database first. Here is the design of the graffiti database you want to appear in your form:

a. What is your name?
b. What is the date?
c. What is the type of graffiti?
   - Ideological
   - Territorial
   - Artistic
d. Does the graffiti require repair?
   - Yes
   - No
e. Where is the graffiti located?

5. Enter your questions.
   It is best practice to enter your questions under label first.

6. Type the name of the field you want to appear in the database.
   a. **name**
   b. **date**
   c. **type**
   d. **repair**
   e. **location**

The last field is the format for the type of answer to the questions.

7. Choose among the predefined options from the pull-down menu.

a. For What is your name?, the answer would be text.
b. For What is the date?, the answer would be a date.
c. For What is the type of graffiti?, the answer would be select_one [list_name].
   In the top tab you want to change [list_name] type.

8. Click Yes to Continue.

One of the greatest concerns when collecting data is data integrity and efficiency. Integrity and efficiency are maintained if the collector is presented with specific choices that maintain the integrity of the attributes. You can arrange this by using the pull-down menus that you choose as a collector. For instance, as the collector in the next step, you will create a collection function that will be either Yes or No.

9. For Does the graffiti require repair?, the answer would be select_one [list_name]. On the top tab, change [list_name] to yes_no.

10. Click Yes to continue.
11. For Where is the graffiti located?, the answer would be a geopoint.

In this way, you have defined two questions to have lists. Lists are shown under choices.

12. Click Choices at the bottom of the page. The single list of choices is Yes or No.
13. Erase the second choice and a set of new choices is added.
The new choices will be types of graffiti: ideological, territorial, or artistic.

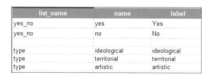

14. At the bottom of the software, scroll left and click Survey.
15. Click File.
16. Click Save.
17. Open Survey123.
18. Open Graffiti.

Prepare your survey for publishing: thumbnail, summary, and description

You are now ready to prepare your survey for publishing. Publishing will send the survey into the cloud where it can be accessed by collectors who can work on the map collaboratively.

1. On the top tab, click Settings.
2. Click the thumbnail, navigate to where a thumbnail image you have prepared is, and insert.
The image must be a .png file (size 100 x 100 pixels) and it must be named the same as the excel worksheet.
3. In this case, name the file **graffiti.png**.
4. Write a brief summary.
5. Write a description.
6. Click the update icon to update your survey.

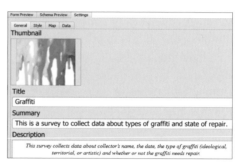

7. Click the publishing icon to send your graffiti survey to the cloud.

You have now completed a basic Survey123 function. There are additional functions that can be added to your survey as well as things to just make your survey look better.

Sharing your survey with members of your organization

1. Go to Survey123 for ArcGIS, survey123.arcgis.com.
2. In the upper right corner, click Sign In to your organizational account.
3. When you see your survey, click the share icon at the bottom of the survey icon.

4. Check Members of My Organization to share with other organizational members.
This allows members of your organization to access your survey on their mobile devices.
5. Click OK.

Next, you will build some additional skills, including setting a survey title, using Images in your survey, and creating subdomains

Set a survey title
The survey title is set from within the spreadsheet.
1. At the bottom of the spreadsheet, click Settings.
2. Change the form_title to Graffiti Collection.
3. Update the spreadsheet.

Use images in your survey
Icon images for selection choices are not only atheistically appealing, but they can also help the collectors make the correct decisions.
1. Search for and choose images that match your selections. Save them as .png files and size them to 100 by 100 pixels.
2. Once again, click the yellow folder icon.
3. Choose media.
4. Paste the images in the media folder.
5. Look for the media image column and type the name of the image in the reference folder that matches the selection. Type the file extension as .png or .jpg.

list_name	name	label	image
yes_no	yes	Yes	
yes_no	no	No	
type	ideological	ideological	ideological.jpg
type	territorial	territorial	territorial.jpg
type	artistic	artistic	artistic.jpg

6. Update your survey.

Create subdomains
You can create conditional logic subdomains that allow the collector to make another choice based on the first choice; in other words, conditional responses. In this instance, subdomains would mean the following.If you choose ideological, you get the following options:
- political: hate groups such as Nazi or ISIS
- racial: race

If you choose territorial, you get the following options:
- tags: gang symbols
- memorials: honors the slain
1. Insert two more rows in the Excel worksheet.
2. Insert the following questions:
   - What type of ideological?
   - What type of territorial?
3. In the name field, type:
   - ideological
   - territorial
4. In the type field, choose one of the following
   - select_one ideological
   - select_one territorial
Remember to add the terms ideological_territorial in the top tag.
5. Scroll to the relevant field and enter the following expressions:
   - ${type}='ideological'
   - ${type}='territorial'

type	name	label	default	readonly	relevant
text	name	What is your name?			
date	date	What is the date?			
select_one type	type	What is the type of graffiti?			
select_one ideological	ideological	What type of ideological?			${type}='ideological'
select_one territorial	territorial	What type of territorial?			${type}='territorial'
select_one yes_no	repair	Does the graffiti require repair?			
geopoint	location	Where is the graffiti located?			

6. Click choices on the bottom tabs, type the choices, and insert the names of any images that you want to display.

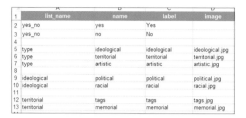

7. When you open this in Survey123, you will see that when you choose hate, the additional question appears.

Lesson 8-6: How bad is the graffiti?

Gathering data for cleanup with your smartphone

Survey123 allows you to use your smartphone or tablet to collect data.

Scenario

Your town council has decided to earmark money for the removal of graffiti. They need to know the location and extent of the graffiti before they can come up with the money for a cleanup plan. They have equipped the Girl Scouts with mobile devices to mark and collect information about the graffiti. They are extremely interested in the type of graffiti that is present.

Build skills in these areas

▸ Downloading the Survey123 app to your mobile device

▸ Downloading the Graffiti survey to your mobile device

▸ Marking a point and collecting the data

▸ Observing collected survey data in a web browser

▸ Observing collected data in ArcGIS Online

What you need

▸ Publisher or Administrator role in an ArcGIS organizational account

▸ Estimated time: 1 hour

▸ Survey123

Download the Survey123 app to your mobile device

1. Go to the App store on your mobile device.
2. Search for and download the Survey123 mobile app.

Access the Graffiti Survey form on a mobile device

1. Click the Survey123 icon on your mobile device.
2. Click the three lines in the upper right corner.
3. Click Download Surveys.

4. Sign in to ArcGIS Online.
5. Click Graffiti and download.

This downloads the survey to your mobile device.

6. After downloading, click Graffiti.
7. Fill out the survey form.
8. Collect several points.
9. Click Send surveys.
This sends your survey points into the cloud.

Access the Graffiti Survey in a web browser

The web browser is where you can share your form with your organization so that others can download. This is also where you can go to review responses to your survey and download your survey data.

1. Go to: http://survey123.arcgis.com/#/.
2. Sign into your organizational account.
3. Click the Graffiti Survey under My Surveys.
4. Click View Results.

The web browser has three tabs: Participation, Data, and Summary.
- Participation tells you the number of surveys submitted and who submitted the survey.
- Data shows the spatial location of the points on a map.
- Summary brings up the results of the survey.

At the bottom of the page, on the right, there are two icons that allow you to save your data as a CSV file or as shapefiles.

5. Click Save.

Access the Graffiti Survey in ArcGIS Online

1. Go to ArcGIS Online and sign in to your organizational account.
2. Click My Content.
3. Click the Survey-Graffiti Folder that you created with Survey123.
4. Click the Graffiti Feature Layer and add to a new map.
5. Click Edit on the top of the page.
6. Click Graffiti in the window pane.
7. Click a point on the map and fill out the survey. Notice there are no subdomains. The subdomains are represented by an additional question.

8. After the editable feature service is added to your map, save your map with the appropriate metadata.

9. In order for the general public to access your map, choose Share with Everyone.
For now, only those in your organization can view the map.

Related Learn ArcGIS resources

Time estimate: 1 hour

Tools(s): Download Survey123, find a survey, collect points in a web browser and in an ArcGIS Online map

Software: Survey123

Audience: Users who want to use the Survey123 app to collect data

This table below offers you a comparison of functionality between Survey123 and Collector.

Comparison of Collector and Survey123

Collector	Mapcentric	Single question response	Visual collection	Relatively hard to change and update	Attachments	Can sync	Captures new data	Can edit existing data
Survey123	Formcentric	Smart form with conditional logic	Nonvisual collection	Easy to change and update	Attachments	Can Sync when prepared correctly	Captures new data	Cannot edit existing data

More online lessons

At the end of chapter 8 in the online version of *The ArcGIS Book* (http://learn.arcgis.com/en/arcgis-book/chapter8/), you'll can explore more about Collector app with the ArcGIS lesson to Manage a Mobile Workforce. Additionally, on the learn.arcgis.com website you'll find a lesson on analyzing the technology habits of your community using Survey 123.

The ArcGIS Book, chapter 8
Questions for reading comprehension, reflection, and discussion

Teachers can use the items in this section as an assignment, an introduction, or an assessment, tailored to the sophistication of learners. Some learners can read all the sections at one time, while others are more comfortable with small segments. The questions and tasks are designed to stimulate thought and discussion.

1. GIS goes where you go and Thought Leader: Jeff Shaner

    a. Discuss the advantages of the mobility of data collection and use an example.

2. What are Collector for ArcGIS and Survey 123?

    a. What is Collector?

    b. What is Survey123?

    c. What are the differences between Collector and Survey 123?

3. Case Study: Gathering data in remote areas

    a.   Is data collection limited only to people and areas that have an Internet connection?

Real-time Dashboards
Monitor live data feeds for specific events and day-to-day operations

Houston, we have a problem. Associated with astronauts running into a crisis on their way to the moon, that line evokes the image of a roomful of teammates in technology trying to help, the image of computers, monitors, and analysts all working together to solve the problem. That image brings to mind the Esri Operations Dashboard app and its role right here on Earth. It is designed to monitor operations by providing a centralized command center for a purpose, toward a shared goal. It can be used to monitor daily logistical operations, such as service deliveries, or critical situational operations like a city's emergency response to a fire. The app also displays the statistics of specific events, such as Olympic competitions or voting returns as they are reported by precinct during a critical election.

An organization can stay on top of its operations by using the app to visualize field data, for example, in customized maps, bar charts, histograms, lists, gauges, graphs, and photo attachments. Live data feeds can be added to the dashboard, and data can be queried or filtered based on attributes such as time. The app is easy to configure and can be shared across your own organization or publically.

In this chapter you will watch example videos of operations dashboards as well as download the app to use on your computer. The chapter will end by providing directions on constructing an Operations Dashboard to track field collections of Wi-Fi strength in Fairfax County, Virginia.

Use the questions at the end of the chapter to support your reading comprehension, reflection, and discussion of the narratives presented in the corresponding chapter 9 of *The ArcGIS Book*.

Introductory activities

Video
Go to video.arcgis.com and search for videos by title.

Raceway Event Protection National Security Demo-Operations
This brief video shows the use of the Operations Dashboard web application for event protection.

Collector and Operations Dashboard for ArcGIS: Monitoring Coastal Security
This brief video demonstrates using Collector for ArcGIS and Operations Dashboard for coastal monitoring missions.

Activity

Operations Dashboard for ArcGIS
http://doc.arcgis.com/en/operations-dashboard/
Download Operations Dashboard: Windows only

Aspects of a real-time GIS

Write a sentence explaining each of the following aspects of a real-time GIS.

1. Acquire real-time data:

2. Perform continuous processing:

3. Communicate the results:

Lesson 9-1: Identify layers
The Boston Marathon is not a cakewalk

And a GIS is not a cake. But information is layered within it in order to make data accessible in distinguishable yet relatable ways. Read the Case study: The 119th Boston Marathon in *The ArcGIS Book*.

Build skills in these areas

▸ Locating Layers

What you need

▸ Account not required

▸ Estimated time: 10 minutes

Lesson 9-2: Gather Wi-Fi strength
Making a map using an editable feature service

You will create a map using an editable feature service of Wi-Fi strength collection in Fairfax County, Virginia. The feature service will provide for a space to record the collector's initials, the date, and the strength of the signal (low, medium, and high).

After the map is created, it will be accessed through the Operations Dashboard app, which will be configured to show the map, legend, description, a bar graph of signal strength, and a bar graph for collector results.

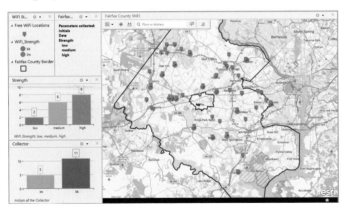

Scenario
Fairfax County Virginia GIS Department received grant funding to examine the current Wi-Fi resources available within their county. After mapping all the known Wi-Fi hotspots found in coffee shops, libraries, schools, and other businesses, they enlist volunteers to use the collaborative map to record Wi-Fi strength. They establish a weekend as the Wi-Fi collection weekend and they build a dashboard to monitor their volunteers.

Build skills in these areas
▸ Downloading Operations Dashboard for ArcGIS
▸ Creating and sharing a map
▸ Creating a Wi-Fi collection dashboard
▸ Customizing the operations view
▸ Assessing the status of the operation

What you need
▸ User, Publisher, or Administrator role in an ArcGIS organizational account
▸ Operations Dashboard (for Windows only)
▸ Estimated time: 1 hour

Download Operations Dashboard for ArcGIS

1. Go to http://doc.arcgis.com/en/operations-dashboard/
2. Download and install the app on your computer (Windows only).

Open the map of Fairfax County WiFi Collection

1. Go to ArcGIS Online and sign into your organizational account.
2. Search for LearnResource.
3. Uncheck within map area.
4. Click the Fairfax County Wifi to open the map.
5. Save to your account with proper metadata.

Open Operations Dashboard and create a new project

1. Open Operations Dashboard for ArcGIS.

2. Click Continue.
3. Sign in to your organizational account.
4. Click Create a New operation view.
5. Click Multidisplay operation view.
6. Click Create.
7. Click the Map widget.
8. Click the Fairfax County WiFi map.

9. Click the three Data Sources you want in your map.

10. Click OK.

Add the Description widget

A widget displays information from a particular data source. Widgets interact with features on the map through feature actions. Feature actions are supported on the map, bar chart, histograms, lists, feature details, and pie charts.

1. On the top of the page, click Widgets.
2. Click Add widget.
3. Choose the Description widget. The Description widget displays a description as static text.

4. Click OK.
5. Position the widget in the upper left corner.

Add Legend widget
1. On the top of the page, click Widgets.
2. Click Add widget.
3. Click Legend widget.
4. Title the Legend WiFi Strength in Fairfax County, VA.
5. Click OK.
6. Position the widget beside the description widget.

Add the Strength bar chart
1. On the top of the page, click Widgets.
2. Click Add widget.
3. Click Bar Chart.
4. Configure with Fairfax County WiFi as the Data Source.
5. Value Field is Strength.
6. Under Axes, check Display labels and grid lines on value axis.
7. Click OK.
8. Position the widget on the lower left of the dashboard.

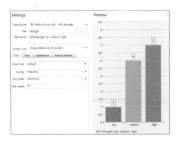

Add the Collector bar chart
1. Repeat steps 1 through 8.
2. For Value Field, choose initials.
3. Position the widget on the lower left of the dashboard.

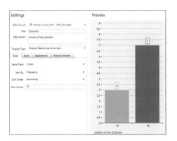

4. Click File and click Save the Operations Dashboard.
5. After you save the dashboard, you can share it with your organization.

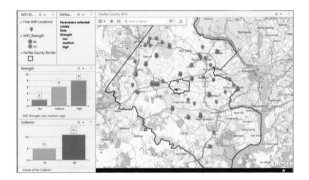

More online lessons
At the end of chapter 9 in the online version of *The ArcGIS Book* (http://learn.arcgis.com/en/arcgis-book/chapter9/), you'll find a Learn ArcGIS lesson to create a real-time dashboard and coordinate an emergency response for emergency vehicles in the City of Redlands, California.

The ArcGIS Book, chapter 9
Questions for reading comprehension, reflection, and discussion

Teachers can use the items in this section as an assignment, an introduction, or an assessment, tailored to the sophistication of learners. Some learners can read all the sections at one time, while others are more comfortable with small segments. The questions and tasks are designed to stimulate thought and discussion.

1. How real-time dashboards are used
   How real is real time?

   a. Describe the aspects of a real-time GIS:

2. Components of a real-time dashboard
   Real-time GIS platform capabilities: Working with real-time data

   a. What is a real-time dashboard composed of basically, and who would you share it with?

3. Examples of real-time data sources
   Case study: The 119th Boston Marathon

   a. Identify some examples of sources of real-time data and how such data could be useful in tracking a specific event:

GIS Is Social

Crowdsource, collaborate, and make your own landmark on the global GIS community

GIS is the prototypical team sport. Contributions from numerous members, hailing from a range of backgrounds and playing many different roles, meld to pursue a shared goal. The GIS game is played at local, state, national, and ultimately, global levels. For example, a state's Audubon society is using GIS to identify endangered habitats with the help of data from its department of environmental management along with bird counts collected by local observers. GIS teams become GIS communities, within which new teams grow as people continue to find each other and connect around shared objectives. Their shared goal, simply put, is to make a difference in the world we live in by using GIS to support informed decision making and problem solving at every level.

Cloud computing and the mobile/app revolution have increased the ability of members of the GIS community to work together and collaborate through data sharing. Just as significantly, technological advances that led to this social age we're in have also expanded the GIS community far beyond the world of professional GIS users. Today, the community potentially includes nearly everyone on the planet. To join in, all you need is a smartphone, tablet, or computer. Whether at work or at home, every time you access data from the cloud, store smartphone photos online, or use a mobile device to report a power outage, you are engaging with and contributing to the global GIS community.

In this chapter you are invited to join a national GIS team and participate in a collaborative project by adding your data to the project map. You will also facilitate crowdsourced collaboration in another initiative by building your own crowdsourcing app.

Introductory activities

Go to www.youtube.com/watch?v=huQpn0D0eK4

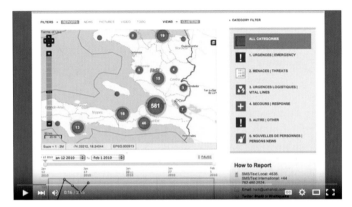

Go to edcommunity.esri.com and search for video by title.

Ushahidi Haiti
In 2010 a devastating earthquake nearly leveled the capital city of Port au Prince in Haiti.
Thousands of Haitians were trapped in the rubble and survivors were in desperate need of medical care, water, and shelter. In the immediate aftermath of the earthquake, first responders and the humanitarian community had little to no information about the situation on the ground. Ushahidi stepped into the information void by using crowdsourced information to create a near real-time map of the city. How did they do it?

GIS Careers: Ned Gardiner - Climate Scientist
GIS has become an integral and essential technology in industries ranging from public safety to public health. This page links to a set of videos that look at the use of GIS in several of these arenas. What types of data are these professionals looking at? How does visualizing and analyzing this data in GIS help them perform their job?

Activity

Go to the Industries page on the Esri website. Select one industry that interests you. Take 10 minutes to explore the information and links about the use of GIS in that industry. In small groups, share what you learned about how GIS is used in the industry you investigated. What are the common denominators and what are the main differences across industries?

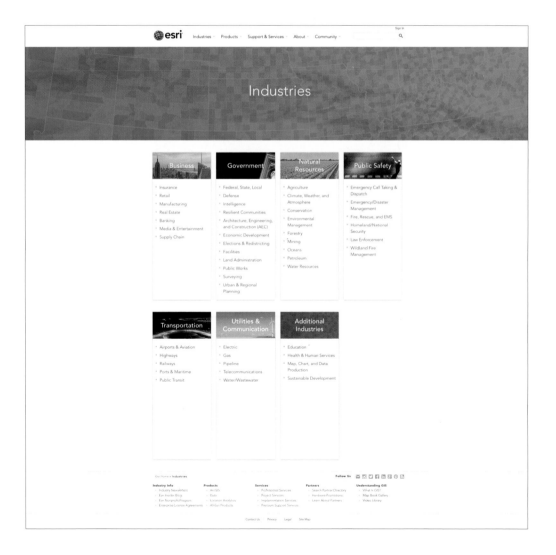

Crowdsourcing

The term crowdsourcing was coined by Jeff Howe in a 2006 Wired article The Rise of Crowdsourcing. We live in a world of perpetual change, according to Howe (Howe, 2006) and crowdsourcing is the only way for 21st century citizens to adapt. Miriam Webster online defines crowdsourcing as the practice of obtaining needed services, ideas, or content by soliciting contributions from a large group of people and especially from the online community. Crowdsourcing is at the heart of the concept known as Citizen Science in which amateur and nonprofessional scientists participate in and contribute data to scientific research. The Ushahidi Crisis Map of Haiti (see video, above) is another example of the power of crowdsourcing.

Wayfinding

Landmarks and wayfinding are two related concepts that lend themselves to use in a GIS through crowdsourcing. In his classic work *The Image of the City*, Kevin Lynch explained that people perceive their city by creating mental maps (city images) they use for what he calls wayfinding. Wayfinding is the use of visual cues in the environment (things you can see) to navigate. These visual cues, or landmarks, are incorporated into our mental map—our city image. Landmarks stored in our mental maps are essential for wayfinding. We use them to move about easily and quickly: to find our way to a store, a doctor's office, or a restaurant.

Landmarks are specific points such as buildings, stores, or gardens that orient us within the community— they tell us where we are. Typically, they are easily identifiable features that stand out against the background—they are hard to miss (a church, a billboard, or a statue). We use landmarks to find our own way or give directions to others so they can find their way. Landmarks reassure us that we're on the right path, tell us how much farther we have to go, and let us know where to turn.

How do you think the landmarks people use for wayfinding in a city differ from those used in a suburb, small town, or rural community? How are the landmarks people use in the West different from those in the South, the Midwest, or New England? There's a way to find out.

Lesson 10-1: Crowdsource your story

Finding Your Way

In this lesson, you will contribute a photograph, a name (including any nicknames), and a description of a landmark in your own community (and the part it plays in wayfinding) to a crowdsourcing story map called *Finding Your Way*.

Scenario

You are a travel writer for your local newspaper in a southwestern city. You are interested in writing an article about regional idiosyncrasies in the description and use of local landmarks. For instance, would you know what it means when a Providence native tells you to take the first exit after the big blue bug? Or a Philadelphian says he'll meet you at the base of the Rocky Steps? To gather data for your article, you post a link on your Facebook page to the Finding Your Way Story Map you created and encourage followers to add distinctive local landmarks found in their own community.

- The Big Blue Bug, also known as Nibbles Woodaway, is the giant metal termite that sits on top of New England Pest Control along I-95 in Providence.
- The Rocky Steps are the 72 stone steps leading to the entrance of the Philadelphia Museum of Art in Philadelphia. The steps are called the Rocky Steps because they appeared in the film "Rocky" and five of its sequels.

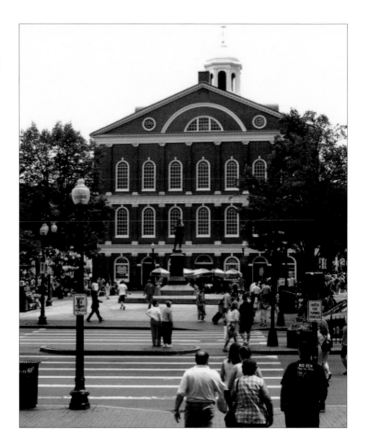

Build skills in these areas

▸ Contributing to a Story Map Crowdsource
▸ Exploring data in a Story Map Crowdsource

What you need

▸ Mobile device or computer with Internet access
▸ Estimated time: 30 minutes

Contribute a landmark to the story
1. Using a smartphone or other mobile device, take a picture of a local landmark that people in your community, city, or neighborhood typically use as a wayfinding landmark (as in the previous scenario). Alternatively, locate a picture of the landmark online being sure to use a source such as Creative Commons to find photos that are either copyright free or for which you can gain permission to use fairly easily.
2. Open the Story Map Crowdsource, Finding Your Way
   a. Go to http://arcgis.com.
   b. Click Search.
   c. Search LearnResource.
   d. Click the Finding Your Way thumbnail
3. Click the "Participate" button on the Crowdsource map or gallery page and sign in.

4. Type in a place name or street address to locate your landmark.
5. Add your photo to the data collection sheet. If you're using a computer you can drag and drop a photo into the box or "Pick a file" to find and upload a photo. If you're using a smartphone you can take a photo or use one from your collection.
6. Fill in the name of your landmark. Include both nicknames (for example, the Rocky Steps) and actual identifiers (for example, stone steps leading to the Philadelphia Museum of Art).
7. In the description box, explain how people in your area typically use this landmark in wayfinding. Is it a place to turn, a proximity indicator (e.g., across the street from, or a quarter mile beyond), and so on?

   What is the history of this landmark? Include any unique anecdotes about the landmark as well. For instance, in Providence, the Big Blue Bug has a bright red light on the end of his nose during December.

   Once you have added your photo and provided the landmark name and description, click submit to add your landmark to the map.

You can add more than one landmark if you choose.

What do landmarks tell you?

1. Explore the landmarks you and others have placed in the Finding Your Way Story Map and look for patterns, similarities, and differences. Consider the following questions as you explore:

   a. What sorts of things in the environment (buildings, billboards, and so on) are used most often as landmarks?

   b. Are landmarks more likely to be man-made elements of the environment (a building) or natural features of the environment (a lake or pond)?

   c. What regional differences do you observe among landmarks?

   d. Are those regional differences related to cultural (Scandinavian heritage), economic (agricultural activities), or physical characteristics (rock formations) of the region?

   e. How do urban, small town, suburban, and rural landmarks differ? How are they similar?

   f. What are landmarks most often used for in wayfinding?

   g. Are there rural/urban or regional differences in the way people use landmarks?

   h. Are you familiar with all the landmarks posted in your community or area?

Lesson 10-2: Celebrate ethnic diversity
Building Story Map Crowdsource Melting Pot

Melting Pot

Crowdsourcing has become an essential component of information gathering and analysis in today's world. Crowdsourcing includes everything from citizens reporting downed trees after a storm to students participating in the scientific investigation of biodiversity called BioBlitz.

If you are a publisher or administrator in an ArcGIS Online organization account, you can design your own crowdsourcing story map by using the Crowdsourcing Story Map Builder. The Builder experience allows you to set the properties of your crowdsourcing app. For example, you can decide whether an addition to the story map is posted immediately after it is created or whether it needs to be approved before posting. Projects can range from local to national, or even global.

In this lesson, you will build a Story Map Crowdsource to reveal ethnic diversity in your community.

Scenario

You are the Director of a nonprofit, community-based organization whose mission is to support immigrant communities through health, educational, and vocational services. You have been invited by the local Rotary Club to give a presentation about ethnic diversity in your city. In preparation for the presentation, you have invited your organization's constituents to contribute to a crowdsourced story map called Melting Pot. Contributors will add a photograph, title, and description of something that reflects the ethnic makeup of their neighborhood such as (but not limited to) nonEnglish signs, ethnic restaurants, and ethnic markets.

Build skills in these areas

▸ Opening the Story Map Crowdsource Builder

▸ Creating a Story Map Crowdsource

▸ Setting properties for the crowdsourcing app

What you need

▸ Administrator or Publisher privileges in an ArcGIS organizational account

▸ Estimated time: 30 minutes

Open the Story Map Crowdsource builder
1. Log in to your ArcGIS organization.
2. Go to the Story Maps page, storymaps.arcgis.com.
3. Click Create Story and select the Story Map Crowdsourcing app to open the app builder.

Create a Story Map Crowdsource
1. Enter app information:
   a. Enter a Title: **Melting Pot.**
   b. Click advanced options to select the folder where the app will be saved.
   c. Click next to create the app.
   d. Add Enter a subtitle: **Crowdsourced map of neighborhood ethnic indicators.**
   e. Change background image of the title screen.

2. Enter directions for Melting Pot data collection:
   Populate this map with photos, locations, and descriptions of things that reflect the ethnic makeup of your neighborhood such as (but not limited to) nonEnglish signs, ethnic restaurants, and ethnic markets.

Congratulations! Your crowdsourcing story map is ready for data entry.

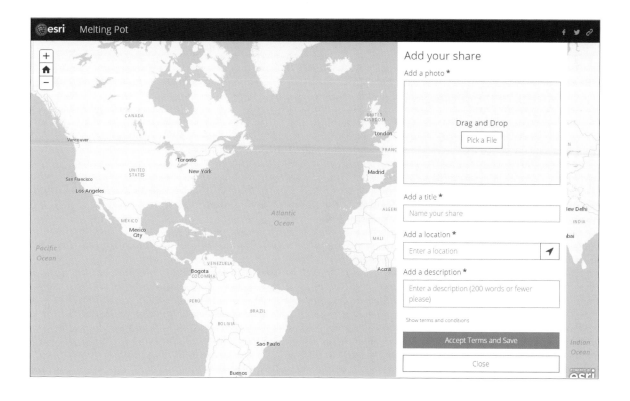

The ArcGIS Book, chapter 10
Questions for reading comprehension, reflection, and discussion

Teachers can use the items in this section as an assignment, an introduction, or an assessment, tailored to the sophistication of learners. Some learners can read all the sections at one time, while others are more comfortable with small segments. The questions and tasks are designed to stimulate thought and discussion.

1. GIS is collaborative: Geography is key for integrating work across communities

   a. How is the concept of open data (see chapter 4) central to understanding the GIS community today.

2. GIS work is a valued profession: Community is vital in GIS

   a. Can you identify some of the unique characteristics of the professional GIS community?

3. ArcGIS for organizations

   a. What are some of the opportunities that an ArcGIS organizational account provide to its members and to the organization overall?

4. Geodesign: Using social engagement in community planning

   a. What are the core elements of geodesign as a planning methodology?

5. Thought Leader: Clint Brown: GIS is participatory

   a. What does Clint Brown mean when he says GIS provides a kind of integration engine?

6. Social GIS and crowdsourcing

   a. What do the three apps shown under the banner Social GIS and crowdsourcing have in common?

7. The rise of community engagement

   a. How do each of the initiatives presented here reflect the concept of community engagement?

8. What is the ConnectED initiative and where does GIS fit in?

   a. ConnectED Goal:

   b. GIS connection:

ConnectED
An education initiative

In response to President Barack Obama's call to help strengthen STEM education through the ConnectED Initiative, Esri President Jack Dangermond announced that Esri will provide a grant to make the ArcGIS system available for free to the more than 100,000 elementary, middle, and high schools in the United States, including public, private, and home schools.

ConnectED is a US government education program developed to prepare K-12 students for digital learning opportunities and future employment. The Initiative sets four goals to establish digital learning in all K-12 schools in the United States

during the next few years. These goals include high-speed connectivity to the Internet, access to affordable mobile devices to facilitate digital learning anytime, anywhere, high-quality software that provides multiple learning opportunities for students, and relevant teacher training to support this effort.

Sites With GIS from Esri and ConnectED

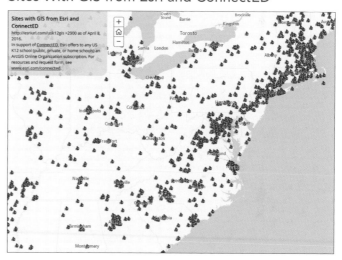

Some regions of the country are engaging aggressively in obtaining an ArcGIS Online account via the Esri ConnectEd effort. **www.esri.com/connected**

GeoInquiries

Your children and your students can engage in learning and applying ArcGIS using this book and the exercises it contains. This is a great way to get your students started with ConnectED.

Quickstart

Set up your ArcGIS Online organization account

▸ **Start the ArcGIS Trial**
From your web browser, visit the ArcGIS Trial page. Fill out your name and email address and click Start Trial.

▸ **Activate the ArcGIS Trial**
Open your email and follow the instructions from the Esri email to activate your ArcGIS Online account. This account, for which you will be the administrator, will allow you and four others to use ArcGIS Online.

▸ **Set up your organization**
There is one more step before your account is activated. Think carefully about your or your organization's short name because this will form the URL for you or your organization (and eventually all your content). Click Save and Continue.

▸ **Get inspired at the Apps page**
To get the Desktop software or other apps, go to the Free Trial page which is always accessible under your account name in ArcGIS.

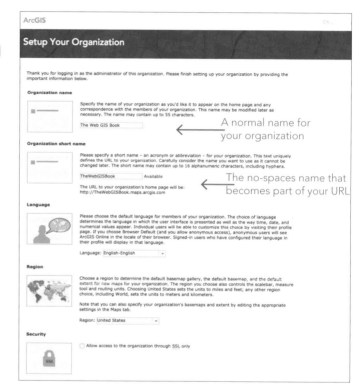

When you set up your trial, pay close attention to these two options. They form your organization's identity as others will see it online.

Resources

Learn.ArcGIS.com

Esri Press

ArcGIS Solutions

GeoInquiries

SpatiaLABS

ArcGIS Marketplace

GeoNet Community

Esri Training

ArcGIS Support

Further reading, books by Esri Press

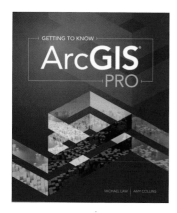

Getting to Know ArcGIS Pro by Michael Law and Amy Collins

Making Spatial Decisions Using GIS and Lidar by Kathryn Keranen and Robert Kolvoord

Making Spatial Decisions Using GIS and Remote Sensing by Kathryn Keranen and Robert Kolvoord

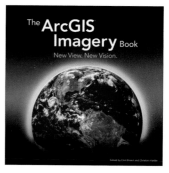

The ArcGIS Imagery Book, edited by Christian Harder and Clint Brown

Getting to Know ArcGIS Pro by Michael Law and Amy Collins

The GIS 20: Essential Skills by Gina Clemmer

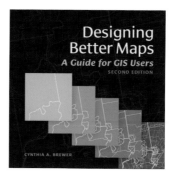

Designing Better Maps by Cynthia Brewer

The GIS Guide to Public Domain Data by Joseph Kerski and Jill Clark

See esri.com/esripress-resources for more books, exercises, data, software, and updates for Esri Press titles.

Acknowledgments

Acknowledgments
Special thanks to Jack Dangermond and Clint Brown for the vision and inspiration that motivated us to write this book. Their enthusiasm and encouragement made it a reality. We are grateful to Charlie Fitzpatrick, Angela Lee, and Joseph Kerski for their excellent support over the years and for their unflagging willingness to collaborate with us as educators. Such guidance has meaningfully influenced the way that we approach students in the classroom, thereby helping us develop the effective teaching methodologies you see here. The assistance of the many members of Esri's software development team has been invaluable. Thanks in particular to Allen Carroll and Steven Sylvia for allowing us to highlight their new crowdsourcing app in this book—we are pleased to be able to feature such new and exciting technology. We appreciate Esri Press' publishing professionalism and cheerful attitude. Finally, we would like to acknowledge Bob Kolvoord and Paul Rittenhouse for their willingness to engage in constant dialogue about geospatial technology education over many, many years.

About the authors

Kathryn Keranen is a retired Fairfax County, Virginia, geosystems teacher. Currently an adjunct instructor at James Madison University, she has taught geospatial technology to both students and instructors for more than 20 years. She co-authored (with Robert Kolvoord) the four-book series *Making Spatial Decisions*(Esri Press).

Lyn Malone is an educational consultant specializing in the classroom application of geospatial technologies. She is the author of *Mapping Our World: GIS Lessons for Educators* (Esri Press) and *Teachers Guide to Community Geography: GIS in Action* (Esri Press).

Credits

Cover

Cover: Map tiles by Stamen Design, under CC BY 3.0. Data by OpenStreetMap, under CC BY SA.

Introduction

Page iiv: *Herndon High School Classroom Photo courtesy of Kathryn Keranen.*

Chapter 1

Page x: *Swisstopo Map*
               Courtesy of Swisstopo.
Page 2:   *Maps We Love: The Power of Maps* Video produced by Esri.
               Access at http://video.esri.com/watch/4285/the-power-of-maps
Page 2:   *Map lovers, feast your eyes on this! 30 Years of Maps* Video produced by Esri.
               Access at http://video.esri.com/search/feast%20your%20eyes
Page 2:   *Geography + technology = "Wow!" What Is GIS?* Video produced by Esri.
               Access at http://video.esri.com/search/wow
Page 3:   *Esri Map Books* Esri.
               Access at http://www.esri.com/mapmuseum
Page 4:   *US States and Cities*
               Map by Kathryn Keranen;
               data sources: ArcUSA, US Census, Esri.
Page 5:   *U.S. Population Change 2000 to 2010* Map by Esri, data sources: Esri, HERE,
               DeLorme, NGA, USGS, HERE. Copyright © 2010 ESRI | Copyright © 2015 Esri.
Page 8:   *Terrain of Swiss Alps (Terrain: Elevation Tinted Hillshade)*
               Map by Esri; data sources: USGS, NGA, NASA, CGIAR, N. Robinson, NCEAS, NLS, OS,
               NMA, Geodatastyrelsen.
Page 10:   *Nepal Earthquake Epicenter Locations*
               Map by Esri; data sources: Esri, DeLorme, NGA, USGS, DeLorme.
Page 14:   *Earthquakes of the World* Map by Esri; data sources: USGS, Esri, HERE, DeLorme, NGA.
Page 19:   *U.S. Population Change 2000 to 2010* Map by Esri, data sources: Esri, HERE,
               DeLorme, NGA, USGS, HERE. Copyright © 2010 ESRI | Copyright © 2015 Esri.
Page 26:   *Live Feed Earthquakes* Map by Kathryn Keranen; data sources: Esri, Delorme, NGA, USGS.

Chapter 2

Page 95:   *Conservation*
Earthstar Geographics | Department of Commerce (DOC), National Oceanic and Atmospheric Administration (NOAA), National Ocean Service (NOS), Office for Coastal Management (OCM) | Source: U.S. Census Bureau | Esri, HERE, DeLorme.

Page 96:   *Fairfax Derecho*
Map by Kathryn Keranen; data sources: ArcUSA and US Census.

Page 102:   *Lincoln Crime*
Map by Kathryn Keranen; data sources: ArcUSA, US Census, Lincoln, NE GIS Department.

Chapter 6

Page 106:   *Rotterdam 3D*
Data sources: City of Rotterdam, Esri Netherlands, Esri.

Page 108:   *Author Web Scenes using ArcGIS Online*
Video produced by Esri.
Access at http://video.arcgis.com/watch/3981/author-web-scenes-using-arcgis-online

Page 109:   *Geography and the Assassination of President Kennedy*
Video produced by Esri.
Access at http://video.arcgis.com/watch/4594/geography-and-the-assassination-of-president-kennedy

Page 109:   *Airflow Globe*
Esri.
Access at http://www.arcgis.com/home/item.html?id=128ba9498cca447ab6ec356b84fee879

Page 109:   *Montreal, Canada Scene*
Esri, Digital Globe, and Microsoft.
Access at https://www.arcgis.com/home/item.html?id=2c1d58da5c244c5dbaadf407c982900a

Page 109:   *High Rise Election*
Map by Kenneth Field, courtesy of Esri.
Access at https://www.arcgis.com/home/item.html?id=f034d08dbde649fe97ed2f3ec5bbf381

Page 109:   *Philadelphia Redevelopment: CityEngine workflow example*
Esri.
Access at http://www.arcgis.com/home/item.html?id=86f88285788a4c53bd3d5dde6b315dfe

Page 110:   *Esri Scene's virtual globe* Esri, HERE, DeLorme, FAO, NOAA, EPA.
Access at http://www.arcgis.com/home/webscene/viewer.html

Page 111:   *View of Syria using Esri's Topographic Basemap*
Esri, HERE, DeLorme, FAO, NOAA, EPA.
Access at http://www.arcgis.com/home/webscene/viewer.html

Page 112:   *View of Malheur Wildlife Refuge Headquarters using Esri's Imagery Basemap*
Earthstar Geographics | Esri, HERE, DeLorme.
Access at http://www.arcgis.com/home/webscene/viewer.html

Chapter 8

Chapter 9

Page 169:   *Fairfax County Wi-fi*
             by Kathryn Keranen; data sources: ArcUSA, US Census, Esri.
Page 170:   *Operations Dashboard for ArcGIS*
             Esri.
             Access at http://doc.arcgis.com/en/operations-dashboard/
Page 173:   *Esri Collector and Operations dashboard apps* Esri.
             *An executive dashboard provides city staff the ability to monitor the status of capital
             improvements, 311 calls, and police patrols* Esri.

Chapter 10

Page 176:   *Ushahidi, Haiti*
             Video produced by Ushahidi.
             Access at https://www.youtube.com/watch?v=huQpn0D0eK4
Page 176:   *GIS Careers: Ned Gardiner - Climate Scientist*
             Video produced by Esri, NOAA
             Access at http://edcommunity.esri.com/Careers/career-videos
Page 177:   *Esri Industries*
             Site by Esri.
             Acess at http://www.esri.com/industries
Page 179:   *Faneuil Hall*
             Faneuil Hall photo courtesy of Isaac Wedin, under CC BY 2.0
             (https://creativecommons.org/licenses/by/2.0/).
Page 180:   *Big Blue Bug*
             Photo courtesy of Big Blue Bug Solutions.
Page 181:   *Rancho Grande: In The Mission district*
             Photo by Phil Whitehouse, under CC BY 2.0 (https://creativecommons.org/licenses/by/2.0/
             legal).
Page 182:   *Story Map Crowdsource Builder*
             Esri.
             Access at http://storymaps.arcgis.com/en/
Page 183:   *Melting Pot Story Map Crowdsource*
             by Stephen Sylvia, Esri; data sources: Esri, HERE, DeLorme, MapmyIndia, ©
             OpenStreetMap contributors, and the GIS user community.
             Access at https://www.arcgis.com/home/item.html?id=6407eb07a3c245bfbbb5cca7e0d65382
Page 185:   *GeoInquiries*
             Site by Esri.
             Access at http://edcommunity.esri.com/Resources/Collections/geoinquiries
Page 185:   Sites with GIS from Esri and ConnectED
             Map by Esri.
Page 187:   Never far from their devices
             Photo by pixdeluxe/Getty Images.